FALLING
STARS

FALLING STARS

A Guide to Meteors and Meteorites

MIKE D. REYNOLDS, Ph.D.

STACKPOLE
BOOKS

Published by
STACKPOLE BOOKS
5067 Ritter Road
Mechanicsburg, PA 17055
www.stackpolebooks.com

Printed in the United States of America

10 9 8 7 6 5 4 3

First edition

Cover design by Wendy Reynolds
On the cover: A meteor shower as seen from space. Original art by Jerry
Armstrong

Library of Congress Cataloging-in-Publication Data
Reynolds, Michael D.
    Falling stars: a guide to meteors and meteorites /Mike D.
Reynolds.—1st ed.
        p. cm.
    Includes bibliographical references and index.
    ISBN 0-8117-2755-6
    1. Meteors. 2. Meteorites. I. Title.
QB741 .R48 2001
523.5'1—dc21
                                                        00-054140

ISBN 978-0-8117-2755-6

To my star,
my wife Debbie

Contents

Acknowledgments

This book has been in progress since 1966. It was November, and as a 12-year-old I was fascinated by space. I had heard about the Leonid meteor shower and how on the morning of November 17 I might just see a lot of falling stars. So I bundled up (my mother always worried about me getting cold while I gazed at the stars) and went outside. What I saw was astounding: *thousands and thousands of meteors;* the sky seemed alive with them. I remember roughly drawing one particularly bright Leonid on a star map. A number of years later I was talking to a good friend who had also seen the 1966 Leonids. We described this one bright Leonid, compared our star maps, and realized that we were talking about the same meteor!

I was very fortunate to have two excellent mentors in my early days of meteor watching: Karl Simmons and the late Richard Sweetsir. I can remember several thirteen-hour December marathon meteor watching sessions with Karl and Richard.

There are several people to thank for their assistance in reviewing this manuscript: Ben Burress, Donna Cox, Ted Cox, Dan Durda, Etta Heber, and Jose Olivarez. Any errors in this book are those solely of the author. Ben Burress assisted with the NASA image research used in *Falling Stars*. Mike Martinez also provided support and research in the area of meteorites; Mike's cataloging of meteorite prices over the past ten years has been an invaluable reference and one I would highly recommend to the individual who is interested in collecting meteorites.

Several friends and associates passed on stories for *Falling Stars*. My sincere thanks to Robert Haag, Mike Martinez, and Blaine Reed for sharing their adventures with all of us.

I was very fortunate to be able to call upon a number of people to assist with the preparation of illustrations for *Falling Stars*. Jerry Armstrong, an amateur astronomer and simply a very talented space artist, captured the vision of this book with his painting for the cover. David Frantz is a creative illustrator; in the midst of his home schooling activities, he prepared illustrations that superbly explain what I attempted to get across with the written word. Chad Thompson also provided several illustrations—his first venture illustrating for a science book. David Sisson and Mike Martinez assisted me with the photography of meteorite and tektite samples; trust me, these rocks are not photogenic! Several individuals allowed me to use their excellent photographs: Bob Drost, Dan Durda, Conrad Jung, Mike Martinez, Jose Olivarez, Blaine Reed, and Carter Roberts. Thanks go to the American Museum of Natural History in New York for photos of the Willamette Meteorite and the 1833 Niagara Falls Leonid Storm painting. Tamara Schwarz assisted me in the scanning of several photographs. I also want to acknowledge the National Aeronautics and Space Administration for the use of several images.

Let me close by thanking three people in particular. Editor Mark Allison and editorial assistant Amy Hixon of Stackpole Books have been encouraging and very supportive of my efforts to write this book. And my sincere thanks go to my wife Debbie! She allowed me to take the time—as well as take over the dining room table—to write *Falling Stars.* In addition, Debbie assisted me in researching meteorites and museums as well as refining some of my crude sketches of meteors and turning them into something more presentable.

So go out and look up! The sky is alive with falling stars. And one of those that you see might be that really bright fireball that produces the next major meteorite fall.

Dr. Mike Reynolds
Oakland, California

1

Falling Stars!

INTRODUCTION

Humans have always been intrigued by the phenomena of the night sky. Occasionally that intrigue was tempered by fear of the apparitions that were seen there, but, on the whole, the night sky was held in reverence. Many stories of bravery, romance, and tragedy have come down to us from people who looked at the stars and saw patterns there. They created the constellations to tell their stories, which were primarily mythical but sometimes historical in nature.

Today we live in a vaster and more comprehensible universe. Yet the universe continues to intrigue us. Interest in astronomy and space exploration is at an all-time high, especially in the United States. People are lining up at public observatories or along telescopes owned by amateur astronomers to glimpse celestial objects, jamming the Internet to view the latest spacecraft images, and flocking to the Kennedy Space Center for the launch of the Space Shuttle. Public meteor shower watches and star parties are still well attended (even at early hours of the morning), and the Smithsonian Institution's National Air and Space Museum in Washington, D.C. is the best-attended museum in the world. Indeed, people today are so intrigued by space and the content of the universe that they would like to touch it, or better yet, own a piece of it. The good news, as you will learn in the pages that follow, is that anyone can do both.

A FEW BASICS

Most of us have looked up at the night sky and seen what is commonly called a falling, or shooting, star. These momentary streaks occur when *meteors,* objects generally ranging from the size of dust particles to fist-size masses, enter the earth's atmosphere at speeds up to 44 miles per second and are ionized (or heated) to incandescence 50 to 75 miles above the earth. Few of these objects survive their encounter with our atmosphere.

What we see here on earth, mostly at night, is a streak of light that lasts about a half a second on the average. Generally speaking, the larger the material that enters the earth's atmosphere, the brighter the meteor. Brighter meteors will occasionally leave a smoke trail behind in their path lasting a few seconds; trails produced by very bright meteors, referred to as *fireballs,* may last minutes. Fireballs that appear to break up or produce sound are called *bolides,* from the Greek word *bolis* meaning missile.

The word "meteor" comes from the Greek word *meteora* or *meteoros.* This term was once used to describe any atmospheric occurrence, such as auroras, lightning, rainbows, and the like. Historically, the altitude at which meteors appeared was a subject of controversy. Some felt that meteors were a local event, like lightning. Others felt that meteors occurred at the same general distances as the stars. What resolved the controversy was the fact that the location of a given meteor in the sky would appear to shift depending from where on the ground the meteor was seen.

This apparent positional shift is referred to as parallax, and can be easily demonstrated. Hold a pencil at arm's length. Close one eye and note the apparent location of the pencil against background objects. Then open that eye and close the other. Note that the pencil seems to shift with respect to the background objects. Now bring the pencil in closer to your nose and repeat the procedure. The apparent shift of the pencil is much greater.

It wasn't until the eighteenth century that the heights of meteors were first calculated, using two observers at different

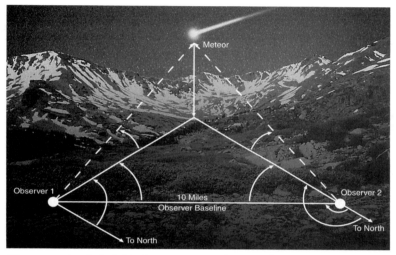

The approximate height of a meteor seen by two observers at two different locations can be determined by using trigonometry. This information can later be used to determine the location of a possible meteorite fall. ILLUSTRATION BY CHAD THOMPSON

locations and parallax. These initial calculations, which showed that meteors were very low in the earth's atmosphere, were wrong for one major reason: The two points of measurement were not far enough apart.

By the end of the eighteenth century, accurate data collected by two German students showed that the altitude of most meteors is between forty-eight and sixty miles above the surface of the earth.

Regular meteor showers occur when the earth, as it orbits the sun, runs into material left behind by sun-orbiting comets. Referred to as dirty snowballs, *comets* are made of ices, which include carbon dioxide and water ices, ammonia, organics, and charcoal and sandlike substances, as well as ions in the gas tail.

Generally speaking, a bright comet appears about once a decade, but many faint comets are discovered each year. Historically they were feared objects, often thought of as bad omens. The Bayeux Tapestry, depicting the Battle of Hastings in 1066, shows

The Bayeux Tapestry. Halley's Comet is at top center. INTERNATIONAL HALLEY WATCH, NASA

Comet Halley over the battle. The comet's appearance actually inspired the forces of William the Conqueror prior to their invasion; the omen was the death of Harold II.

Comets are believed to be from two general areas in our solar system: the Oort Cloud and the Kuiper Belt. Oort Cloud comets are at the far extremes of the solar system, believed to be 20,000 to 100,000 astronomical units from the sun (one astronomical unit, or AU, is the distance from the earth to the sun, or 93 million miles). There may be trillions of icy objects in the Oort Cloud, some of which are pulled from the cloud by occasional passing stars, changing their orbits enough to send a few of these icy bodies toward the sun.

A closer region of icy bodies, the Kuiper Belt, lies just beyond the gas planets and within the plane of the solar system. These bodies were probably formed from the solar nebula, far enough from the Jovian planets to have survived. Even though direct observation of the Kuiper Belt has not, at this point, been possible, enough bodies have been discovered to lend credence to its existence. Some of the outer solar system objects, such as Chiron,

Neptune's moon Triton, Pluto's moon Charon, and Pluto itself, might actually be or have been Kuiper Belt objects. (Thus a controversy has erupted regarding Pluto and its status as a planet.)

The time it takes a comet to go around the sun in its orbit is its period. Short-period comets, which come from the Kuiper Belt icy bodies, have periods of 200 years or less and tend to lie relatively close to the plane of the solar system (within 30 degrees). The shortest period comet currently known is Encke's Comet, with a period of 3.3 years. Halley's Comet has a period of 76 years. The orbits of short-period comets are often narrow and sometimes described as cigar-shaped.

Long-period comets are those with periods greater than 200 years. As a general rule, comets from the Oort Cloud have very long periods. Comet Kohoutek, visible in 1973–74, has a period of 80,000 years. Long-period comets have orbits that are elliptical, parabolic, or even hyperbolic in shape and can approach the sun from all directions.

The major parts of a comet are its head, or coma, a nucleus of material hidden within the coma, and the tail. The coma of a comet is a huge cloud of dust and gas, some being many times the size of the earth. In 1957 Comet Mrkos exhibited a coma that was estimated to be fifteen times the earth's diameter. The Soviet *Vega* spacecraft, during the 1986 rendezvous with Halley, encountered dust particles nearly 190,000 miles from the comet's nucleus! Spectral analysis shows that the gas of the coma is made of carbon, hydrogen, nitrogen, and oxygen atoms in ion and molecular form. Hydrogen molecules appear to form a "cloud" that surrounds the coma/nucleus; this cloud is larger than the sun and appears to be coming from the breakdown of water molecules within a comet's nucleus.

The nucleus of a comet can be up to tens of miles across. From multiple spacecraft flybys of Comet Halley in 1986, we know the nucleus can be very dark in color (thus the term "dirty snowballs") as well as very irregular in shape. The nucleus cannot be seen except by direct spacecraft observation; it is well hidden within the coma.

Comets can have two types of tails: gas and dust. The gas

tail, called a type I tail, appears fairly straight with streaks and irregularities. Gas tails are made up of ionized atoms that have been released from the nucleus. Gas tails are most intriguing; they can disassociate from the comet nucleus, which then forms a new gas tail.

Dust tails, or type II tails, appear rather smooth, with little or no visible features. The dust is very small—about one micron, or a millionth of a meter, in size—and simply reflects sunlight. Whereas gas tails are blown relatively straight by the solar wind, dust tails are usually arced due to the differential velocity between the coma/nucleus and the dust tail.

Dust tails pose a real threat to approaching spacecraft. In fact, the European Space Agency's *Giotto* spacecraft, which made the closest flyby of Comet Halley in 1986, was damaged due to dust grain impacts.

The type of tail formed is particular to each comet. In 1996 Comet Hyakutake exhibited a spectacular type I gas tail, but no dust tail. In 1997 Comet Hale-Bopp, one of the most spectacular comets of the twentieth century, exhibited both a spectacular

Comet Hyakutake, 1996. The comet exhibited a beautiful gas tail. PHOTO BY CONRAD JUNG

Comet Hale-Bopp, 1997. The comet's yellow dust tail and blue gas tail are easily visible. PHOTO BY CONRAD JUNG

blue gas tail and a yellow dust tail. But research has shown that not all comets have tails; in fact, there is evidence that the majority of comets do not have visible tails. These comets have apparently exhausted their ices.

As a comet approaches the sun, the comet's nucleus—the concentration of the cometary mass—begins to heat up and "outgas." Gas and dust released by the comet's nucleus first envelope the nucleus, then are swept "behind" the comet. This sweeping effect is due to solar wind, fast-moving ions and atoms from the sun's corona that are moving outward through the solar system. Because of the solar wind, the tail of a comet, if it forms a tail, always points away from the sun. As these comets travel through the inner solar system, they leave a trail of material behind, somewhat like the proverbial trail of crumbs.

The mostly dustlike material left behind by comets as they orbit the sun is the source of several annual meteor showers. Even though cometary dust and meteoritic debris cause them, meteor showers are generally named for the constellation from which they seem to emanate. Some examples are the Leonids

(Leo), Geminids (Gemini), and Perseids (Perseus). The "id" ending means "sister" or "children of." Occasionally a shower *radiant*—the point in the sky from which the meteors seem to originate—is defined not only by the constellation, but by a specific star within the constellation, such as the Eta Aquarids. Just to keep things interesting, there are exceptions to the rules. For example, when the Draconid meteor shower "storms" (hundreds or thousands of meteors each hour), it is named the Giacobinids after the Comet Giacobini-Zinner that causes the shower.

Not all meteors are produced by meteor showers. Random, or *sporadic meteors* can be seen on any night. It is likely that the first meteor you ever saw was one of these. Although we generally do not see shower or sporadic meteors during the daytime, occasionally an extremely bright meteor will be visible during daylight under clear sky conditions.

OTHER SPACE DEBRIS

Meteors very rarely impact the earth. Most are vaporized fifty to seventy-five miles up. But the larger source particles of very bright sporadic meteors that survive their encounter with the atmosphere and impact the earth are called *meteorites.*

Material in space not associated with a comet prior to entering the earth's atmosphere is referred to as a *meteoroid.* Meteoroids can be made up of various types of materials, often material left over from the formation of the solar system. (Meteorite composition classes are discussed in chapter 5.)

Other types of bodies that can strike the earth include comets and asteroids. *Asteroids* (the word means "starlike") are usually about a kilometer across. They are larger masses of rock and metal that predominately orbit the sun between the orbits of Mars and Jupiter, but a few are in orbits that cross the inner planets (Mercury, Venus, Earth, and Mars). Impacts from comets and asteroids are catastrophic events; it is theorized that the end of the dinosaur era sixty-five million years ago was due to a large asteroid or comet impact in the Yucatán Peninsula, Mexico. The object, estimated to have been six to seven miles across, formed a crater 125 miles in diameter, called the Chicxulub impact feature.

Recent research has shown that the collision produced a crater nearly eight miles deep and ejected 12,000 cubic miles of rock, dirt, and debris into the earth's atmosphere. The impact also produced fires, acid rain, and destructive waves. The Chicxulub impact was so great that it actually changed the shape of the base of the earth's crust at a depth of twenty-two miles.

The first asteroid discovered was Ceres, observed in 1801 by the Sicilian monk Giuseppe Piazzi, who believed it to be a planet orbiting between Mars and Jupiter. Astronomers had long theorized that a planet should be found at that distance because of a theory known as Bode's Law (which was eventually proven to be incorrect). Bode's Law stated that the distances of the planets from the sun could be predicted by a mathematical equation. However, Ceres was small compared to the other planets (a little over five hundred miles in diameter).

Eighteen years later another object was discovered in approximately the same orbit as Ceres. A number of additional objects were discovered in the following years, but none are larger than Ceres and only three of the asteroids are more than 250 miles in diameter.

After the initial discovery of multiple asteroids in the space between Mars and Jupiter, it was theorized that a single planet once orbited there. According to the theory, the planet broke up, most likely due to the gravitational pull of its gigantic neighbor, Jupiter. Today's astronomers believe that the asteroid belt is simply made up of material left over from the formation of the solar system. Astronomers also theorize that there were initially about seventy smaller bodies that broke into the thousands and thousands of asteroids that now form the belt.

Astronomers have classified asteroids based on their orbits. One group, the Apollo objects, are asteroids that have earth-crossing orbits. The gravitational pull of Jupiter affects the orbits of these asteroids, making them quite dangerous as their orbits shift.

Close-up exploration of asteroids had to wait until recently. On its way to the planet Jupiter, the spacecraft *Galileo* made two all-important flybys of the asteroids Ida and Gaspra. Images sent

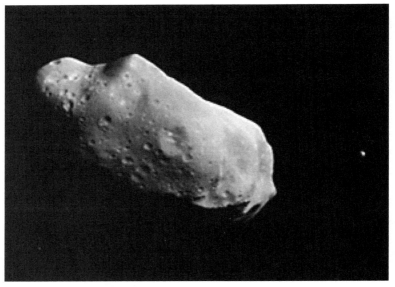

The heavily cratered asteroid Ida, thirty-five miles long, and its satellite Dactyl, to the right. This image, taken by the Galileo *spacecraft, was the first to prove that asteroids can have satellites.* PHOTO COURTESY OF THE JET PROPULSION LABORATORY, NASA

back by *Galileo* revealed irregular-shaped bodies that had been battered by impacts. And in the case of Ida there was an additional surprise: an asteroid moon.

In 1996 the United States launched a spacecraft specifically to orbit and study an asteroid. The Near Earth Asteroid Rendezvous, or NEAR, arrived at the asteroid Eros in February 2000. Data from NEAR indicates that Eros, a twenty-one-mile-long asteroid, is an irregular-shaped solid body, showing many impact features and no moons or debris surrounding the asteroid. Eros, apparently a very old asteroid, passes close to, but never crosses the earth's orbit.

What do we know about those asteroids that pass near the earth? In the early 1930s an asteroid was discovered passing within ten million miles of the earth. The first of many, these objects became known as the near earth asteroids, or NEAs. Most were once members of the main asteroid belt but were ejected by Jupiter. There is also a possibility that some NEAs are dead comets. NEAs are considered dangerous to the earth if they pass within three million miles.

In 1973 the Palomar Observatory Asteroid Survey was initiated. The survey found just under one hundred NEAs. In the 1980s, the use of charged-couple devices, or CCDs, instead of photographic plates made the detection of asteroids, including NEAs, much quicker. Lowell Observatory's Anderson Mesa site is home to a twenty-four-inch Schmidt telescope with CCD imaging; astronomers at Lowell discover an average of one new NEA *each night*. And most recently, the use of a CCD coupled with a super-computer to process data resulted in the discovery of thousands of previously unknown asteroids in less than a year.

Not all of these recently discovered asteroids are NEAs, or earth killer asteroids. It is believed that approximately 2,500 asteroids fit into the earth killer category; however, it is estimated that only ten percent have been discovered. The odds of an earth killer asteroid impact, similar to the one that might have brought

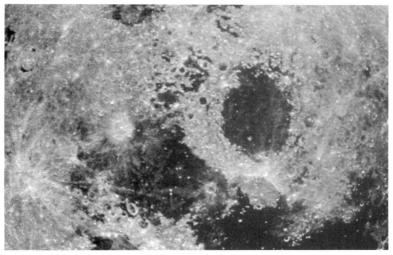

The earth's cratered moon as seen by Apollo 11, *July 1969.* PHOTO COURTESY OF NASA

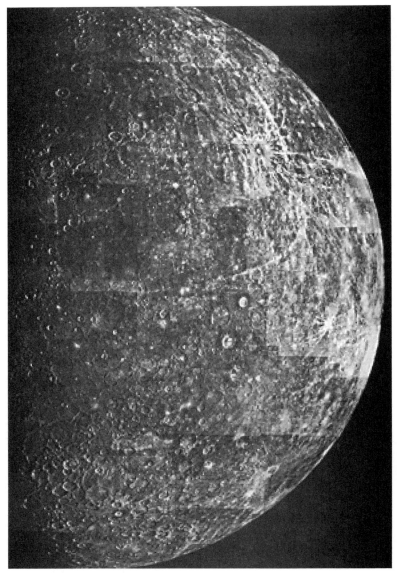

Many planets and satellites in the solar system exhibit cratering due to impacts. The planet Mercury exhibits multiple craters, which makes it look much like the earth's moon. MARINER 10 IMAGE, NASA

an end to the dinosaurs, are thought to be less than one in several hundred thousand. It is estimated that a major earth impact by an asteroid of a half-mile in diameter or larger occurs once every 200,000 years.

Scars of impacts can be easily seen on our planetary neighbors in space. Our moon is scarred with thousands of impact craters. Fortunately not all impacts are catastrophic. In fact, we can enjoy and benefit from the occasional meteorites that fall to earth and are recovered each year.

The casual or regular meteor observer can contribute to the recovery of a freshly fallen meteorite by recording specific information about the parent fireball's path. Thus the efforts of the sky watcher and the ardent meteorite collector go hand in hand.

2

Meteor Watching

INTRODUCTION

Meteor watching might be considered by some to be the most romantic of all the astronomical endeavors. It requires the least equipment and, for the most part, the least knowledge of the night skies. Meteor watching is a great way to introduce the novice to the constellations, to star identification and the stellar magnitude sequence, and to the relationship of mathematics to astronomy. Meteor watching is something that one can do alone, or as part of a group. Serious meteor watchers can contribute significantly to the field of meteoritics, including the recovery of meteorites. And observations during a very active meteor shower or one producing bright fireballs are quite thrilling. So let's get started!

OBSERVING EQUIPMENT

For basic meteor watching, the observer should have the following:

- *Chaise lounge or lawn chair.* The type of chair that works best is the lawn chair with an adjustable back that allows the observer to fully recline. The adjustable back allows for the best possible observing angle. Regular chairs are not recommended; the observer will strain his/her neck looking up and around. Lying directly on the ground is also not recommended, even in a sleeping bag. And standing up for an extended period of time is not an option. Some observers also use pillows or cushions on their lawn chairs for additional comfort.

- *Sleeping bag or blanket.* Even during warm summer nights a light sleeping bag is useful for padding the chaise lounge for comfort, helping to keep the mosquitoes away, or for warmth during cool mornings.
- *Insect repellent.* Use your favorite repellent or skin lotion. Mosquitoes can quickly ruin a night of meteor watching.
- *Food and drink.* A thermos of coffee or a favorite beverage and some snacks can make meteor observing sessions more enjoyable, especially those after midnight.
- *Recording supplies.* Bring a clipboard, pen or pencil, meteor-observing forms, and red-filtered flashlight. (All flashlights must have either a red filter over the front lens or a red diode as their light source. Some observers are now using a green diode flashlight. Unfiltered light will ruin your night vision, and, if you or someone else is attempting meteor photography, will "fog" the film.) Some observers prefer to use a small tape recorder to record individual meteor observations and then transcribe the data at a later date.
- *Binoculars.* A pair of 7x50 binoculars is handy for looking at meteor "trains." Lenses smaller than fifty millimeters (50 mm) do not gather sufficient light for astronomical observing. A magnification of seven power is about right for hand-held binoculars—a higher magnification and the image will be difficult to keep still. Larger, high-powered binoculars, like the modern 11x80s or 25x100s, require a tripod and mount for steady viewing and are probably not necessary for the casual meteor watcher. However, the larger binoculars are excellent for watching meteors (called *telescopic meteor watching*) when properly mounted.
- *Dress.* The quickest way to end a meteor watching session is to be inappropriately dressed for cold weather. Dress

in layers, which gives you the opportunity to peel off clothes as needed. The winter months bring some excellent meteor showers, but also cold temperatures.

RECORDING DATA

The simplest form of meteor data collection is the hourly count. A tally counter or clicker, which can be purchased at an office supply store, is an easy way to make hourly counts. Tally counters are especially useful when shower activity is high.

Some observers like to keep their meteor tally by making tick marks on paper—but remember you are doing this in the dark and might need to look away from the sky while you are making a tally.

A tape recorder can also be used to record meteor counts. Make certain the batteries are fresh, and that the tape works. Familiarize yourself with the use of the tape recorder before going out to the meteor watch. You will be using the recorder in the dark, so practice using it in this manner beforehand. (Note that some recorders might not work under very cold conditions.)

If several people are observing together, their hourly counts should not be pooled. Hourly counts are specific to each observer. Likewise, you should not include in your count a meteor another friend has observed that you missed.

During very active meteor showers, and especially around a shower's apparent peak activity, it is useful to make meteor counts every fifteen minutes, instead of once an hour. This will eventually allow astronomers to more accurately determine the time of the meteor shower's peak.

Two tally counters will be needed if you are counting both sporadic and shower meteors. Be prepared to record your counts on an hourly basis.

Many observers record more specific data on each meteor, including the following:

- *Time when the meteor is seen.* Observations to the minute are usually sufficient. Use an accurate time source, such as radio station WWV or WWVH, broadcast on shortwave radio by the National Bureau of Standards. This will assure accuracy to the minute.

- *Meteor class.* Indicate whether the meteor is a shower or sporadic meteor—an indication of its source. Occasionally there may be two or more meteor showers visible on the same night, so code the meteors to indicate which shower they belong to.
- *Brightness.* You can make an estimate of a meteor's brightness by knowing the brightness, or *magnitudes* of surrounding stars. Make a comparison and then record an estimate of the meteor's brightness to the nearest half magnitude, if possible. (New observers have a tendency to overestimate the brightness of a meteor, mostly due to the element of surprise the meteor presents.)

The visual magnitude scale, referred to by astronomers as the apparent magnitude scale, was originally used to differentiate between the stars according to their appearance from the earth. The brighter the object, the smaller the magnitude. For example, a 1st magnitude star is 2.512 times brighter than a second magnitude star and 100 times brighter than a 6th magnitude star. Usually the faintest stars visible to the unaided human eye are around 6th magnitude. Sirius, the brightest star in the sky (with the exception of the sun) is 970 times brighter than a 6th magnitude star. The scale developed by Hipparchus was originally devised for 1st through 6th magnitude stars but had to be extended after the invention of the telescope, which increased our ability to see fainter objects. Sirius is –1.5 magnitude and the star Vega in Lyra is 0.0 magnitude. The planet Venus can be as bright as –4.4 magnitude; the full moon is –12.5 magnitude. The sun is a brilliant –26.7! The limit for the Hubble Space Telescope is around magnitude 30. Meteors rivaling the brightness of Venus are seen, but fireballs that rival the brightness of the full moon are very rare.

- *Color.* Meteors appear in colors other than white. Red, orange, yellow, green, and bluish meteors, as well as basic white, are seen. Colors often indicate the meteor's chemical makeup, as well as changes in temperature the meteoritic particle is undergoing.
- *Length.* Estimate how far the meteor traveled, in degrees, by using the distances between stars for comparison. For example, look at the bowl of the Big Dipper in the constellation Ursa Major. The distance between the two stars marking the rim of the bowl, Merak and Dubhe, is five degrees. Extend your hand to arm's length and see how many fingers fit between Merak to Dubhe; this is your personal marker for five degrees, which you can apply to other parts of the sky.
- *Train.* Some meteors leave a smoke trail, or "train." Estimate how long (in seconds) the train is visible, as well as its length in degrees. Binoculars can be used for a more detailed inspection of the train; look for structure and how it drifts in the atmosphere. If you time how long you see a train through binoculars, also time how long you can observe the train visually. If overall meteor activity is slow, time should be taken to sketch the train. Some observers

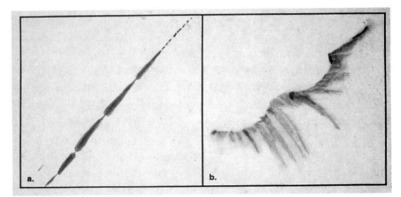

In-field meteor sketch observations. Left: a bright fireball. Right: a meteor train. SKETCHES BY DEBBIE REYNOLDS

have been fortunate enough to photograph or even video-tape a train.

- *Notes.* Include here any unusual observations of the meteor.
- *Serial number.* Some observers keep an annual running tally of meteors they have observed using a six-digit format: 99-0001, 99-0002. The first two digits represent the last two digits of the calendar year, and the four digits after the dash represent the running annual meteor count.

A sample *Visual Meteor Observing Form* can be found on page 20. There are several organizations listed in Appendix B that are interested in visual meteor data.

CONDITIONS FOR METEOR WATCHING

The number of meteors you see each hour can be influenced by several factors, not the least of which are your local conditions. Bright streetlights and other artificial lights will severely hamper your efforts to see meteors. Try to find a park or remote spot that is relatively dark and safe (remember you might be meteor watching at 3 A.M.). If you are observing from a forest and have a limited view of the sky, you will miss seeing a number of low meteors, thus a low or flat horizon is preferred, if possible. Meteor watching from the top of a mountain or the desert is not necessary; however, take advantage of these locations if possible.

Whenever possible, avoid meteor observing while the moon is up, especially if it is between first and third quarter. Sometimes this isn't possible, for a meteor shower might peak while the moon is rising or setting.

Haze or smog will also cut down on the number of meteors you see. Often observers will include in their meteor watching reports an indication of the quality of the sky, discerned by observing the faintest star seen at the overhead point (or the zenith). Also record the percentage of the sky visible behind passing clouds and above a less-than-favorable horizon.

Meteor showers and the occasional sporadic meteor are often best seen after midnight because we are then facing the direction in which meteoroids are colliding with the earth's atmosphere at

VISUAL METEOR OBSERVING FORM

Observer _____ Place _____

Date of Observation ___/___/___

Session began at _____ Session ended at _____

Sky Conditions at the beginning of the Session _____

Sky Conditions at the end of the Session _____

Additional Notes or Comments _____

No.	Time	Class	Mag.	Color	Length	Train	Notes	Serial Number

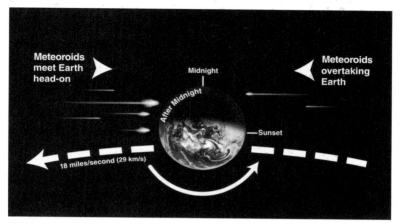

As the earth rotates, meteoroids are more likely to hit the earth's atmosphere on the forward facing part of the earth. ILLUSTRATION BY DAVID FRANTZ

high speed. After midnight, the earth is literally running into the material, acting as a big dustpan (before midnight, the material has to catch up with the earth).

One can expect to see about six or seven sporadic meteors per hour before midnight under good conditions. This number can almost double after midnight.

METEOR SHOWERS

During its orbit around the sun, the earth occasionally collides with streams of dust and other material. As discussed in chapter 1, comets leave behind most of this dust and debris as they orbit the sun, although at least one shower stream has been associated with an asteroid (though some believe the object is actually a dead comet).

The collision of a stream with the earth produces multiple meteors, thus a meteor shower. Meteors seem to radiate out from a general area because of perspective, but actually the stream meteoroids are traveling through space in parallel orbits. As the earth enters the stream, we view the meteors in much the same

Even though we know railroad tracks are parallel, they appear to converge to a single point, much the same as shower meteors appear to radiate from a single point. ILLUSTRATION BY DAVID FRANTZ, PHOTO BY BOB DROST

way as looking down a railroad track. Even though we know the two tracks are parallel, they appear to converge at a distant point.

Tracking the radiant, which tends to drift from night to night, is possible by carefully plotting the location of the shower meteors on a star chart over several nights of observations, then tracing their paths back to a general area in space. The Lyrids are a great shower for tracking the radiant; a three-night period of Lyrid observations will suffice.

When showers are listed or announced by the media, an estimated rate for the number of meteors you can expect to see is given (for example, sixty Perseids an hour). This is the *zenithal hourly rate,* or ZHR. The ZHR shower activity rate is based on the number of meteors you could see under ideal conditions, when the shower's radiant is directly overhead, in the zenith. The ZHR also presumes clear skies and no moonlight or other artificial light to interfere with observing. In some cases, especially with showers of a short peak duration, the shower radiant might even be below the horizon at the peak of the shower, so that few or even no meteors are seen. If you are recording shower and meteor data, you should note the altitude of the radiant from your site at the beginning of each hour.

One of the best showers year to year is the August Perseids because of the number of meteors the observer can see. Unless the moon—or poor weather locally—interferes, you can count on fifty to one hundred meteors each hour. The Perseids are best after midnight, and actually closer to dawn. The best times to observe meteor showers vary from shower to shower; details on major and minor meteor showers are given in chapter 3.

When watching a meteor shower with a group, each person can face in a different direction to better cover the entire sky. If

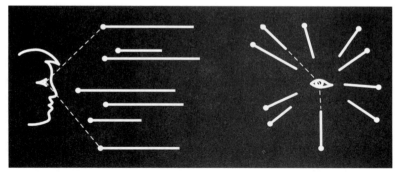

Meteors may appear to be coming straight at you, but they are actually radiating from one area. ILLUSTRATION BY DAVID FRANTZ

you are part of a group and recording data on each meteor, make certain you include in your data what part of the sky you are facing.

Rare treats, *meteor storms,* often very short in duration, occur when the earth passes through a dense swarm of meteoric material. If conditions are right (overhead radiant, clear weather, and little or no moonlight), a meteor storm is a stunning and memorable event, described as observing snow flurries at night.

FIREBALLS, BOLIDES, AND SPACE DEBRIS

Occasionally an observer will be treated to a very bright and characteristically slow meteor called a *fireball,* loosely classified as such if it is brighter than the planet Venus. They can last up to several seconds as they make their fiery entry through the earth's atmosphere. Fireballs can be bright enough to cast shadows; often this is how the observer will first notice them. Fireballs can actually rival the full moon—and on very rare occasions the sun—in brightness. They also have a tendency to leave a train. These trains can also last longer than average meteor trains and some fireballs appear to shed sparks as they enter the atmosphere. Bright fireballs can often be seen during the daytime, and occasionally someone will film or videotape such an event. In August 1972, a woman in Wyoming who was using an 8-mm zoom movie camera caught an extremely bright daylight fireball on twenty-six seconds of film.

With the advent of the space age, spacecraft and space debris reentering the earth's atmosphere have produced human-made fireballs. Often these fireballs are very bright and colorful, producing brilliant and long-lasting trains. Occasionally NASA and other space agencies will know of a satellite's reentry and will issue an advisory on the expected date and time of reentry.

Historically, mid-April to early May have produced very bright, prominent fireballs with no known radiant. Their flights are of long duration and often produce spectacular trains. Though the direction of the April-May fireballs in the sky has varied over the years, look for them toward the southeast, in the evening sky. Occasionally a fireball will explode and split into pieces during its fiery flight, creating objects called *bolides*. Some bolides may "deliver" one or more meteorites that impact the earth.

Fireballs and bolides may produce sound that is audible for several seconds after the object is seen. A loud cracking noise might be heard in the area directly underneath the fireball's path and for a short distance around this area. A low-pitched rumble or thundering noise might be heard over a much larger area and for an extended period of time. Bolides producing "whooshing" noises have also been reported.

OTHER METEOR NOTES

Another unusual celestial object you might see from time to time is a *point meteor*. This is a meteor that is heading straight toward you. Instead of seeing a flash across the sky, you would see a quickly brightening stationary point. Just hope that the point meteor isn't too bright; it could lead to an impact near your location. Point meteors are rare, and often difficult to recognize though they have been seen in numbers during meteor storms.

Occasionally observers will see a meteor while they are looking through a telescope or pair of binoculars. Telescopic meteors look no different from those seen with the naked eye, but will appear brighter; you will also have the opportunity of viewing the meteor's train if one is produced. Most important, however, is the ability to see fainter meteors with the telescope or binoculars than with the naked eye. There is evidence of wholly telescopic meteor showers, that is, showers that are only visible through a telescope or binoculars because the shower's meteors are all faint.

Telescopic meteors are more difficult to catch since a telescope or even a pair of binoculars significantly reduces the observer's field of view. Usually a pair of binoculars, a short-focus/wide-field telescope, or a comet-seeker telescope is chosen for telescopic meteor work; low magnifications are preferred.

As a meteor enters the earth's atmosphere, it leaves a brief ionized trail, which makes possible the detection of meteors using radio waves. Radio meteor "observing" does not require sophisticated equipment for making counts; an FM-band radio is all you

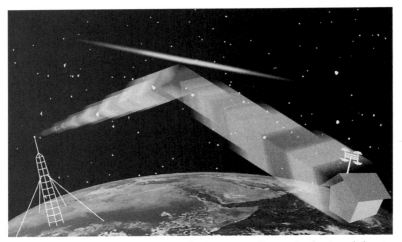

The momentary audible bounce of a distant radio station's signal due to a meteor's ionized trail. ILLUSTRATION BY CHAD THOMPSON

need. Tune the radio to a distant FM station, one that lies over your local horizon (approximately forty to fifty miles away). The signal you receive will be very weak or not audible at all. When a meteor passes between the FM transmitter and the "observer," the ionized trail created by the meteor will deflect the station's signal downward, creating a burst of strong (normal), audible signal. The observer then counts the number of bursts (each lasting from a second up to a minute or more if the meteoroid was large).

PHOTOGRAPHING METEORS

Capturing a meteor on film is a challenge for both the novice and the experienced astrophotographer alike. It's not because it takes elaborate equipment or special techniques; it's because photographing a meteor is more the result of a stroke of good luck than photographic skill or fancy equipment.

Meteor photography is scientifically useful and can provide data on shower activity such as the range of shower meteor magnitudes and a count of meteors for a specific area of the sky; it also can pinpoint the shower's radiant.

The equipment necessary for meteor photography is quite simple. It includes:

- A camera capable of taking time exposure photographs (usually a 35-mm or larger format)
- Cable release for time exposures
- Tripod
- Film

The technique for general meteor photography is similar to star trail photography. If you were to set a camera on a tripod and take a thirty-second or longer exposure, the photograph would show the stars as lines or trails, because of the earth's rotation. The same technique is used for simple meteor photography. The key here is luck—having your camera pointed in the right direction at the right time.

Today's autofocus cameras are not a good choice for meteor photography. Taking long duration exposures of the type required for meteor photography will drain the camera's batteries. A 35-mm manual camera body, capable of taking time or "bulb" exposures, will do the job. A cable release with a lock, attached to the camera body, will allow for time exposures.

Bright meteor in Sagittarius. PHOTO BY CARTER ROBERTS

Generally speaking, it is best to use a normal lens, with focus on infinity, and the aperture or f/stop set to the smallest number. A telephoto lens will reduce the field of view and your chances of catching a meteor. Some astrophotographers have had good luck with a fisheye or a shorter focal length lens ("normal" lenses usually have a focal length of around 50 to 55 mm). A shorter focal length allows coverage of a larger area of the sky, increasing the chances of photographing a meteor in its entirety. The lens must be of high quality—thus usually rather expensive. For example, a new, good quality manual camera body costs around $200-$250, and a good quality 16-mm lens is three to four times that amount! Although some meteor photographers have had good results photographing the sky using a convex mirror or even a highly polished automobile hubcap in the place of a high-quality lens, distortions and reduced brightness have been a concern.

A good, solid tripod is required since you will be taking time exposures. Some meteor photographers prefer a mount like those found on many telescopes, with a clock drive device for tracking star motion. Such a mount is nice, but not necessary.

The type of film—prints or slides—depends upon your personal preference. Each has its own advantages. The speed of the film is important; the faster the ISO/ASA, the better your chances of catching a faint meteor. Grainy film—due to faster speed—is not much of an issue anymore. However, be aware of the fact that the faster films are more likely to "fog" due to excess moonlight or artificial light.

The technique is simple. Properly load the film in the camera. Set the tripod in place and attach the camera. Again, remember to check the focus (infinity), aperture (lowest f/stop), and exposure (time or bulb setting). Now you are ready to photograph.

If you are attempting to photograph shower meteors, it is best not to center the camera on the radiant. Instead, center your camera on an area up to forty-five degrees from the radiant. Also try to avoid the horizon and its thicker layer of air; the greater the altitude the better.

Photographs can be taken up to five to ten minutes in duration, depending on the film speed and amount of external light. Longer photographs are possible, but film fogging, as previously mentioned, is an issue. In addition, if you are using meteor photography to pinpoint a meteor shower's radiant, excessive exposure time will make that effort more difficult if you are using the star trail method. This is because over the exposure time the radiant will rotate, and it is more difficult to determine the radiant from photographed meteors.

One of the aggravations of meteor photography is artificial light. This usually comes from three sources: cars, streetlights, and airplanes. The headlights from cars—other observers arriving or leaving or those out on an early morning drive—will quickly ruin a photograph. Be prepared to carefully cover your lens with an opaque cloth or cardboard if you see a car approaching—or better yet, end the exposure. By ending the exposure you don't chance bumping the camera. Also beware of other observers who have nonfiltered flashlights—these can expose your film in a way you don't want!

Airplanes represent another problem to meteor photographers. One or two photos of star trails with an airplane's lights passing through the field is okay. Airplanes do not fog or interfere with the photograph or lessen in any way the chances of catching a meteor. They simply detract from the aesthetics of the photograph.

In addition to lights, you also need to watch out for moisture forming on the camera lens, especially if you live in an area prone to high humidity. Solutions to this problem include using a lens shade to shield the lens or a portable hair dryer to dry off any moisture. Never attempt to dry the lens by rubbing it with cloth. Many a lens has been scratched when someone tried to dry it with a shirt tail.

After setting the camera on "T" (for time exposure) or "B" (for bulb-style cable releases), carefully begin the exposure. To avoid vibrations, some meteor photographers use a piece of black cardboard in front of the lens as a manual shutter when starting and ending the photograph.

Some meteor photographers use multiple cameras to cover a larger area of the sky, thus increasing their chances of photographing a meteor. This is an excellent plan, but you must be very organized—remember you are photographing in the dark.

After visually observing shower activity for some time, you might get an idea of where the most meteor activity is occurring. However, it often seems that when you move your camera, the area in the sky where the camera was previously pointing picks up in meteor activity!

Keep a log of your photographs. Include the following in the log:

- General site and date information
- Camera equipment information
- Data about each exposure, including the time you started and ended each photograph, the general area of the sky where the camera was pointed, and any suspected meteor activity in the area during the exposure.

When having your film developed, make certain the film developer knows the exposed film is of astronomical images. This is important for two reasons: (1) If print film, the developer might think that it was incorrectly exposed because the prints are all of lines (star trails and possibly a meteor or two); (2) if slide film, the person mounting the slides might not be able to tell where each slide begins and therefore may cut the film incorrectly (a solution to this is to take one normal exposure at the beginning and the end of the roll, in addition to informing the developer that the film is of astronomical images).

Other types of more advanced meteor photography are possible. You may use a propeller-like device in front of the lens. As the propeller or moving shutter rotates, it will "chop" the meteor trail into segments. Knowing the speed of the propeller in number of revolutions per minute will allow you to determine the duration of the meteor. Another advanced project is spectroscopic meteor photography, which can provide information about the meteor's chemical composition.

Sometimes a foreground object, such as a tree or a mountain, will make for an interesting meteor photograph. One photograph taken by the author in 1977 of the city of Boulder, Colorado, with "star trails" above had a pleasant surprise: a Perseid meteor! Occasionally meteor photography succeeds best by chance.

3

Meteor Showers

A number of major and minor meteor showers can be viewed each year. As you are planning which shower to observe, remember that a number of conditions, including the phase and location of the moon relative to the shower, will influence what you see. New minor showers are periodically discovered. If you note activity from an area of the sky that seems unusual, make a note of it and collect as much data as possible. Plots of meteors from a potentially new shower are very useful in determining the radiant. Some minor showers are not seen year to year; the value in observing these shower meteors is to determine activity (if any) and the strength of the particular shower. A number of minor showers with little or no recent activity are not included in this listing; you should consult other references or, even better, meteor-observing web pages, for these showers and updated information.

The following overview includes the shower's name, date of maximum, constellation of the radiant, and zenithal hourly rate (ZHR), if known, as well as specific comments on each shower. The average speed of the meteors, in miles per second (mi/s) and kilometers per second (km/s), is given for many of the listed showers as a reference point; note the great range of speeds, from 9 mi/s for Tau Herculid meteors to 43 mi/s for the better-known Leonids. As you review individual shower data, you might ask

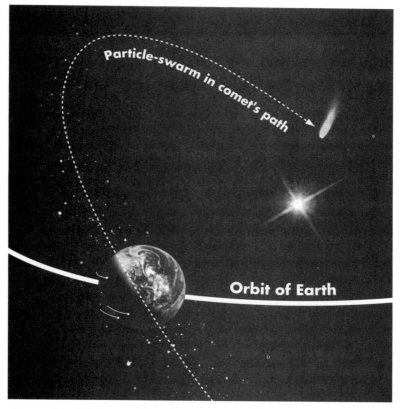

As comets orbit the sun, they leave a stream of particles. If the earth's orbit intersects this stream, a meteor shower will occur. ILLUSTRATION BY DAVID FRANTZ

why some seemingly high ZHR showers are not considered major meteor showers. The primary reason is shower reliability; the ZHR rates listed are maximums observed under ideal conditions. Historical notes of significance are included on some showers. An overview table of all the showers listed is included at the end of the chapter.

The showers overviewed do not include any Southern Hemisphere meteor showers (unless meteors are observable from the southern United States), daytime meteor showers (unless meteors

Shower meteors appear to radiate from a specific area in the sky.

ILLUSTRATION BY DAVID FRANTZ, PHOTO BY DR. MIKE REYNOLDS

from the shower are visible just after sunset or prior to sunrise), or radar meteor showers (those meteor showers primarily detectable by radar). A couple of telescopic meteor showers are included; often more meteors are seen with these showers using optical aid, but naked eye meteors, albeit a small number, are also visible.

JANUARY

Quadrantids (January 3 and 4) in the constellation Bootes with a ZHR of forty-five to two hundred meteors per hour. The interval of this shower is very short, from December 28 to January 7, but it has an output potentially as great as the August Perseid shower. You will need to check an astronomical publication such as *Astronomy* or *Sky & Telescope* to determine the best time to view the shower. The radiant rises around midnight, with viewing possible until dawn. The radiant is never very high for observers in the United States; the farther north you are, the better. Quadrantid meteors are relatively fast at about 25 mi/s (41 km/s) and are bluish in color.

The name Quadrantids comes from an old name for the group of stars between the Big Dipper and the constellation Hercules, the Quadrans Muralis.

Rho Geminids (January 8) in the constellation Gemini. A minor shower, with indications of a secondary maximum around January 21.

Delta Cancrids (January 17) just west of the star cluster Praesepe in the constellation Cancer with a ZHR of perhaps only four meteors per hour. This is a minor shower of fast-moving meteors that can be seen from December 14 to February 14.

Alpha Leonids (last week of January) in the constellation Leo. The Alpha Leonids are an excellent example of a telescopic meteor shower, though some members are visible to the naked eye, and at least one reference lists a ZHR of ten meteors per hour. Alpha Leonids can be seen from January 13 to February 13, with some reported activity as early as the end of December. This is a good minor shower to attempt telescopic meteor watching, or to try to confirm visually on one of those cold, clear winter nights.

FEBRUARY

Aurigids (February 5 to February 10) in the constellation Auriga near the bright star Capella with a ZHR of two meteors per hour. Aurigids can be seen from January 31 through February 23 and are slow, bluish, faint meteors, though an occasional fireball is seen.

Delta Leonids (February 22) in the constellation Leo in the lion's back with a ZHR of three to five meteors per hour. Though a minor shower, the Delta Leonids have a long period from early February to as late as March 19. The radiant is well placed for observing, being overhead around midnight, after which the activity should pick up. These are slow meteors with average speeds of 15 mi/s (24 km/s).

MARCH

Pi Virginids (between March 3 and March 9) in the constellation Virgo with a ZHR from two to five meteors per hour. The Pi Virginids can be seen from February 13 to April 8 and is one of several minor meteor showers than can be seen in the same general area of the sky, making for worthwhile spring meteor watching.

Beta Leonids (around March 20) in the constellation Leo with a ZHR of three to four meteors per hour. Meteors from this minor shower can be seen from February 14 to April 25; the date of the maximum is not precisely known.

Eta Virginids (around March 18) in the constellation Virgo with a ZHR of one to two meteors per hour. The Eta Virginids can be seen from February 24 to March 27, and like the Beta Leonids, the date of maximum is not precisely known.

Theta Virginids (around March 20) in the constellation Virgo with a ZHR of one to three meteors per hour. Theta Virginids can be seen from March 10 to April 21.

APRIL

Alpha Virginids (April 7 to April 18) in the constellation Virgo with a ZHR of five to ten meteors per hour. These slow meteors with a long maximum peak can be seen from March 10 to May 6. You can count on a good display from this reliable minor shower.

Gamma Virginids (April 14 and 15) in the constellation Virgo with a ZHR of up to five meteors per hour. Gamma Virginids can be seen from April 5 to April 21 and add to the already complex activity of minor showers from the constellation.

Lyrids (April 21 and 22) on the border of the constellations Lyra and Hercules with a ZHR of ten to twenty meteors per hour. The Lyrids are historically famous as one of the oldest meteor showers, first recorded by Chinese astronomers in 687 B.C. At one time this shower, produced by Comet Thatcher (last visible in 1861), was quite strong. The Lyrids can exhibit showers of over one hundred meteors per hour, one of which was observed in 1982. Lyrids are usually bright and relatively fast, with speeds of about

29 mi/s (48 km/s). Watch for Lyrids from April 16 to April 25, beginning about 11:00 P.M. local time, looking overhead. As night progresses toward morning, look fifty to sixty degrees above the horizon, lying with your feet to the south, west, or north (it is not a good idea to look directly at the radiant). Observers have noted bright fireballs from the Lyrids a few days after maximum.

April-May fireballs (April 30) with no known radiant. The April-May fireballs produce a wonderful but sporadic display in the evening skies from about April 15 through the first week of May. The author has recorded up to four brilliant April-May fireballs with trains in one evening. They have been known to produce meteorites.

MAY

Eta Aquarids (May 5) in the constellation Aquarius with a ZHR of twenty or more meteors per hour. The best spring meteor shower, the Eta Aquarids are remnants of the rather famous Comet Halley. This shower was first recorded by the Chinese in A.D. 401. Eta Aquarids can be seen from around April 21 to May 12 and the shower maximum can vary. The radiant is found in the water jar of Aquarius and moves daily a little to the northeast. Expect fast, yellow meteors moving at speeds of 40 mi/s (65 km/s), many of which will leave trains. This shower is best observed just before dawn when rates can approach fifty meteors per hour. Rates are higher the farther south you are located, but the Eta Aquarid radiant never gets very high in the sky before dawn, so your observing time is limited. Bright fireballs from the Eta Aquarids can be seen from May 9 to May 11.

May Librids (May 6) in the constellation Libra with a ZHR of two to six meteors per hour. You can see this minor shower from May 1 to May 9.

Northern May Ophiuchids (May 18 and 19) in the constellation Ophiuchus with a ZHR of two to three meteors per hour. Meteors from this minor shower can be seen from April 8 to June 16, making the Northern May Ophiuchids a long-duration shower. There is also a southern stream visible for a shorter period of time, April 21 to June 4, with a maximum occurring between May 13 and May 18. This shower produces about one or two meteors per hour, so it is possible to see three to five meteors per hour coming from the constellation of Ophiuchus. Some references refer to the latter stream as the *Southern May Ophiuchids,* or the *Xi Ophiuchids.*

JUNE

Tau Herculids (June 9) in the constellation Hercules with a ZHR of one to two meteors per hour. The Tau Herculids can be seen for a period of about a month, from May 19 to June 19. These meteors are very slow—about 9 mi/s (15 km/s)—and very faint. This is an evening shower, with the Tau Herculid radiant nearly overhead at 10:00 P.M. local time. Tau Herculid meteors are faint, so you will have to observe from a dark sky to see activity from this shower. Reports have put rates as high as fifteen meteors per hour, but you will probably see substantially less than this.

Theta Ophiuchids (June 10) in the constellation Ophiuchus with a ZHR of ten meteors per hour. Meteors from this shower can be seen from May 21 to June 16, with the maximum occurring over a five-day period centered on June 10. A minor shower that can be bright in appearance, Theta Ophiuchids are best observed around midnight and from southern latitudes. This shower has been referred to by other names: *Alpha Scorpiids, Scorpiids-Sagittariids,* and *Delta Sagittariids.*

June Lyrids (June 15) in the constellation Lyra near the star Vega with a ZHR of eight meteors per hour. The June Lyrids can be seen from June 10 to June 21. This shower is related to the May Lyrids, and yields mostly faint meteors; about one third of them leave trains. The best observing period is between 10:00 P.M. and dawn; the radiant is overhead at about 1:00 A.M. local time. Amateur astronomers discovered the June Lyrids in 1966.

Pi Sagittariids (June 18) in the constellation Sagittarius with a ZHR of five meteors per hour. Meteors from this stream can be seen from June 1 to July 15. This shower may also be listed as the *Scorpiids-Sagittariids.*

Ophiuchids (June 20) in the constellation Ophiuchus with a ZHR of about six meteors per hour. The second shower in the constellation Ophiuchus in the month of June, the Ophiuchids can be seen all night long, but, like the Theta Ophiuchids, are best observed around midnight. Past observations have reported a ZHR potentially approaching twenty meteors per hour at maximum. Even though the average Ophiuchid meteor is somewhat faint, a number of bright meteors and even fireballs have been reported and confirmed.

June Bootids (June 28) in the constellation Bootes with a ZHR of one to two meteors per hour. This faint shower can be seen from June 27 to July 5, occasionally producing a bright meteor. The June Bootids produced good displays in 1916, 1921, and 1927.

June Scutids (June 27) in the constellation Scutum with a ZHR of about two to four meteors per hour. Meteors from this minor shower can be seen from June 2 to July 29.

JULY

Alpha Lyrids (July 14) in the constellation Lyra with a ZHR of one to two meteors per hour. The Alpha Lyrids, primarily a telescopic meteor shower, can be seen from July 9 through July 20. These relatively faint meteors appear to be white in color. At the time of the shower's discovery in 1958, the Alpha Lyrids were producing rates of eighteen or more meteors each hour as viewed through binoculars.

Southern Delta Aquarids (July 29) in the constellation Aquarius with a ZHR of fifteen to twenty meteors per hour. The Southern Delta Aquarids, one of two distinct meteor showers in the Delta Aquarid stream, have been a reliable show. Best observed after midnight, members of this shower can be seen from July 14 through August 18. Rates of ten or more bright meteors per hour (with average speeds of about 26 mi/s (42 km/s) have been observed for as long as a week, leading many observers to report that the Southern Delta Aquarids do not seem to have a sharp peak. A number of observers have reported nice trains associated with Southern Delta Aquarid meteors.

This shower is often listed as the *Delta Aquarids* even though it is distinctly two showers: *Southern Delta Aquarids* and *Northern*

Delta Aquarids. The Southern Delta Aquarids are the more active of the two streams. With careful plotting of shower meteors you should note two radiants. Northern Delta Aquarids peak about the same time as the Perseids on August 13.

AUGUST

Alpha Capricornids (August 1) in the constellation Capricorn with a ZHR of six to fourteen meteors per hour. This shower is visible from July 15 to September 11. Alpha Capricornids are, on the average, brighter than the more famous Perseids.

Southern Iota Aquarids (August 6) in the constellation Aquarius with a ZHR of seven to eight meteors per hour. Look for Southern Iota Aquarids from July 1 through September 18, at speeds of 21 mi/s (34 km/s).

Upsilon Pegasids (August 12) in the constellation of Pegasus with a variable ZHR; estimated rates are one to two an hour, though a few observers have reported higher. A Florida amateur astronomer first observed the Upsilon Pegasids in July 1975 while he was observing early Perseid members. Upsilon Pegasids appear white to yellow in color and do not leave trains. The speeds of Upsilon Pegasids are around 31 mi/s (51 km/s), a little slower than the Perseids.

More observations of this shower are needed. Make certain you do not confuse Perseids with Upsilon Pegasids.

Perseids (August 12 and 13) in the constellation Perseus with a minimum ZHR of fifty to sixty meteors per hour. The Perseids, visible from July 17 until August 24, are the most reliable of modern showers, well-placed in the sky for northern hemisphere observers during the summer months. The main show is after midnight, on the morning of August 12. After midnight, the constellation Perseus becomes more prominent, and shower rates escalate. Perseus is nearly overhead at sunrise, so expect rates to increase until the light of dawn begins to interfere. Observers have reported Perseid rates well in excess of two hundred meteors per hour! It has been estimated that approximately forty-five percent of all Perseids leave trains.

Perseids are generally bright, yellow meteors, with speeds

around 36 mi/s (59 km/s) and average brightness of 2.3 magnitude. Expect to see several fireballs, some bolides, and up to forty-five percent of the meteors exhibiting excellent trains. This crowd-pleasing shower is created by remnants of Comet Swift-Tuttle.

The only problem with the Perseids is distinguishing them from coinciding showers. August 12 is not the best time to begin learning how to plot meteors and radiants or collect individual meteor data. Just enjoy the show; try to make individual shower and sporadic counts if possible.

In some parts of the world, the Perseids are called the Saint Laurence Tears in deference to the martyrdom of Saint Laurence on August 12, 258. This shower was recorded as early as A.D. 36, and annual records of the Perseids have been kept every year since 714. The recognition for discovering the Perseids annual appearance is usually given to Quételet in Brussels, who, in 1835, reported the annual August shower is from the constellation Perseus.

Northern Delta Aquarids (August 13) in the constellation Aquarius with a ZHR of up to ten meteors per hour.

Alpha Ursid Majorids (August 13 and 14) in the constellation Ursa Major with a ZHR of four meteors per hour. This shower, which competes with a number of others, may produce more meteors telescopically. Additional observations, both visually and telescopically, are needed.

Kappa Cygnids (August 18) in the constellation Cygnus with a varying ZHR of up to six meteors per hour. This minor shower is visible from July 26 to September 1. Moving at speeds of 16 mi/s (26 km/s), Kappa Cygnids are characteristically white-yellow or bluish white meteors, some of which leave spectacular trains. Numerous fireballs have also been reported. This shower was first seen by Chinese astronomers in 1042.

Northern Iota Aquarids (August 25) in the constellation Aquarius. This minor shower, visible from August 11 to September 10, is often compared with its companion shower, the Southern Iota Aquarids. Providing a good opportunity to plot meteors and differentiate between shower radiants (unless you get caught up in the Perseid frenzy!). Expect Northern Iota Aquarid speeds of 19 mi/s (31 km/s), a little slower than the Southern Iota Aquarids.

SEPTEMBER

Alpha Aurigids (September 1) in the constellation Auriga with a ZHR of up to nine meteors per hour. Alpha Aurigids can be seen from August 25 to September 6. Two outbursts of this shower, producing over thirty meteors per hour, were recorded in 1935 and 1986.

Gamma Aquarids (September 7) in the constellation Aquarius with a ZHR of one to four meteors per hour. Members of this minor shower can be seen from September 1 to September 14.

Southern Piscids (September 11) in the constellation Pisces with a ZHR of about five meteors per hour. A weak, minor shower, the Southern Piscids, also known simply as the *Piscids,* can be seen from August 12 to October 7, exhibiting relatively slow speeds of 16 mi/s (26 km/s). An even weaker *Northern Piscid* stream may be active about the same time. It appears not to be associated with the Southern Piscids.

OCTOBER

Eta Cetids (first week of October) in the constellation Cetus. The Eta Cetids, visible from September 20 to November 2, are a very weak shower, at best. Occasionally fireballs from this stream are seen.

Delta Aurigids (between October 6 and October 15) in the constellation Auriga. Members from this stream can be observed from September 22 to October 23. Observations of the Delta Aurigids are needed.

Autumn Arietids (around October 8) in the constellation Aries with a ZHR of three to five meteors per hour. Members of this minor shower can be seen from September 7 to October 27. The Autumn Arietids might be associated with the *Southern Taurids.*

Draconids (October 9) in the constellation Draco with a ZHR range of one meteor per hour to a meteor storm. The Draconids, produced by Comet Giacobini-Zinner, are not seen every year.

Their short duration is from October 6 to October 10; the peak occurs during an approximately four-hour window on October 9. Expect Draconid speeds of 12 mi/s (20 km/s).

When the earth passes directly through the parent comet's stream, the results can approach six thousand or more meteors an hour, as were seen in the 1933 and 1946 Giacobinid storms. If Comet Giacobini-Zinner has just made a pass near the earth, rates of several hundred an hour can be observed. Draco, being a circumpolar constellation, is visible from dusk to dawn, and it is best to begin observing as soon as it is dark. Observe the Draconids lying with your feet toward the east to northeast.

Epsilon Geminids (October 19) in the constellation Gemini with a ZHR of one to two meteors per hour. Members from this minor shower can be seen from October 10 to October 27.

Orionids (October 21) in the constellation Orion, near his club, with a ZHR of twenty meteors per hour. The Orionid shower is visible from as early as October 15 through October 29. This shower is part of the stream left behind by Halley's Comet and is an example of an old meteoroid stream. Orion rises about 11:00 P.M. local time; however, Orionids are best seen in the morning sky. Orion is near its highest point at about 4:00 A.M. local time. Orionids are very fast meteors, exhibiting speeds of 41 mi/s (66 km/s). Many shower members are faint, requiring dark skies and an attentive observer. Color varies but will be mostly white, and reports state that twenty percent of all Orionids produce a train. When observing the Orionids, look straight up or about sixty degrees above the southwestern horizon.

Southern Taurids (October 30 to November 7) in the constellation Taurus. The Southern and Northern Taurids (see next entry) have a combined ZHR of seven meteors per hour. The Taurids are produced by the famous periodic Comet Encke, and are best viewed after midnight. Seen as early as September 17, Southern Taurids exhibit rather slow speeds of 17 mi/s (28 km/s) and can produce bright meteors and fireballs with spectacular trains throughout November. You might see this shower listed as the Taurids when consulting other references.

NOVEMBER

Northern Taurids (November 4 to November 7) in the constellation Taurus. Northern Taurids can be seen from October 12 to December 2, with speeds of 18 mi/s (29 km/s).

Andromedids (around November 14) in the constellation Andromeda with a ZHR of about five meteors per hour. Meteors from this weak shower can be seen from September 25 to December 6. This shower produced many thousands of meteors per hour in 1872 and 1885. Sometimes the Andromedids are referred to as the Bielids, after the rather famous Comet Biela (which split into two comets in 1846, most likely causing the Bielid storms of 1872 and 1885).

Leonids (November 17) in the constellation Leo with a ZHR of ten to fifteen-plus meteors per hour. The reliable Leonid shower is best seen in the morning skies between November 14 and November 21. Leonid meteors are bright and very fast, exhibiting speeds of 43 mi/s (71 km/s).

If the earth crosses a dense section of the stream left behind by its parent comet, Temple-Tuttle, a meteor storm will occur, and the Leonids have provided the most famous. Arabic nations declared A.D. 902 "the year of the stars" because of the Leonids. A Chinese text records the Leonid activity in A.D. 931, stating "many stars flew, crossing each other" and "many stars flew and fell."

In 1833, the sky came alive with Leonids. People on the North American east coast saw a spectacular 50,000 to 200,000 meteors per hour, describing them as being like snow or rain. Several North American Native American tribes, including the Sioux and the Maricopa, made records of the 1833 Leonids.

Every thirty-three or thirty-four years the earth passes through a very dense section of the Leonid stream, producing a storm like that seen in 1833. In 1866 another storm was seen, but the pull of Jupiter on the stream may have caused poor showings in 1899 and 1932. However, thirty-four years after the disappointing 1932 show, the Leonids again stormed, producing 150,000 meteors an hour over the midwestern United States on the morning of November 16, 1966.

Astronomers and meteor enthusiasts, as well as the general

Painting depicting the 1833 Great Leonid Storm as it appeared over Niagara Falls. COURTESY DEPT. OF LIBRARY SERVICES, AMERICAN MUSEUM OF NATURAL HISTORY, NEG. NO. 119942

public, were looking forward to the possibility of another great Leonid storm in either 1998 or 1999. A spectacular series of very bright Leonids as well as fireballs were seen in 1998, but no apparent storm. In 1999 a storm apparently did occur over Europe, but was brief in duration and in some cases not seen due to poor weather.

Predictions for the future hold great promise, with some astronomers predicting thousands of meteors each hour at the peak through 2002—if you are standing in the right place on the earth. Strong peaks have been projected for the Americas in 2001 and 2002.

Lie in your chaise lounge or lawn chair with your feet to the east. Don't look directly at Leo or the radiant; look about thirty degrees to the west of Leo. Leonids will appear to come from the Lion's "sickle." The radiant rises after midnight local time, but you can begin observing as early as 11:30 P.M.

November or Alpha Monocerotids (November 21) in the constellation Monoceros with a ZHR of a few meteors per hour. Meteors from this shower can be seen from November 13 through December 2. The radiant is highest at about 1:00 A.M. local time. The November Monocerotids are noteworthy because every ten years the shower produces strong activity of around one hundred meteors per hour. Such activity was last seen and confirmed in 1995.

DECEMBER

Delta Arietids (December 8 and 9) in the constellation Aries with a ZHR of only one meteor per hour. Meteors from this weak shower can be seen from December 8 to January 2.

Chi Orionids (December 10) in the constellation Orion with a ZHR of five meteors per hour. The Chi Orionids are actually two showers (Northern and Southern), each producing similar rates at about the same period of time. Chi Orionids are usually bright; about one in seven leaves a train.

December Monocerotids (December 11) in the constellation Monoceros with a ZHR of one to two per hour. The December Monocerotids can be seen from November 9 to December 18. This shower is an excellent candidate for telescopic/binocular viewing. Additional observations would help to confirm December Monocerotid shower activity.

Geminids (December 14) in the constellation of Gemini with a ZHR of fifty to one hundred meteors per hour. Many seasoned meteor watchers believe the Geminid shower is better than the August Perseids. You can expect to see Geminids from December 6 to December 19. The Geminids peak very quickly, and the morning of the maximum is by far the best time to observe. Gemini is at its highest point near midnight, so the evening of the 13 and the morning of the 14 are the best times to watch, with most of the activity occurring after midnight on the 14. Geminids are bright, yellow meteors, the average magnitude being reported around 2.4. They have speeds of 21 mi/s (35 km/s), and only about one in twenty Geminids produces a train.

There has been some debate over the years as to the parent of the Geminids. It had been suggested that the asteroid Icarus produced the stream. In 1983 an asteroid-like object named 3200 Phaethon was discovered in the same orbit as the Geminid stream. Observations have pointed to the possibility that Phaethon is actually a comet that is no longer active.

Coma Berenicids (around December 19 to December 29) in the constellation Coma Bernices. A weak shower, Coma Berenicids have no definite peak, with meteors being seen from December 12 through January 23. Coma Berenicids are fast meteors, exhibiting speeds of 40 mi/s (65 km/s).

Ursids (December 22 and 23) in the constellation Ursa Minor with a ZHR of ten to twenty meteors per hour. Ursids can be seen all evening from December 17 through December 25, since the radiant is in a circumpolar constellation that never sets. Usually faint meteors, Ursids have speeds of 21 mi/s (34 km/s). In 1945, the Ursids produced a ZHR of over one hundred meteors per hour. In 1986, increased hourly rates for the Ursids were also reported.

Zeta Aurigids (December 31) in the constellation Auriga. This minor meteor shower, visible from December 11 to January 21, is primarily a telescopic (or radar) shower; a few members may be visible to the naked eye. A northern branch of the shower is also visible, with maximum occurring January 2. Zeta Aurigids are generally slow-moving meteors.

SUMMARY OF ANNUAL METEOR SHOWERS*

Meteor Shower	Active Period	Peak	ZHR Max
Quadrantids	December 28–January 7	January 3–4	45–200
Rho Geminids	December 28–January 28	January 8 (and 21?)	
Delta Cancrids	December 14–February 14	January 17	up to 4
Alpha Leonids	January 13–February 13	January 24–31	
Aurigids	January 31–February 23	February 5–10	
Delta Leonids	Early February–March 19	February 22	3–5
Pi Virginids	February 13–April 8	March 3–9	2–5
Beta Leonids	February 14–April 25	Around March 20	3–4
Eta Virginids	February 24–March 27	Around March 18	1–2
Theta Virginids	March 10–April 21	Around March 20	1–3
Alpha Virginids	March 10–May 6	April 7–18	5–10
Gamma Virginids	April 5–April 21	April 14–15	up to 5
Lyrids	April 16–April 25	April 21/22	10–20
April–May fireballs	April 15–May 7	April 30 (varies)	varies
Eta Aquarids	April 21–May 12	May 5	20+
May Librids	May 1–May 9	May 6	2–6
Northern May Ophiuchids	April 8–June 16	May 18–19	2–3
Tau Herculids	May 19–June 19	June 9	1–2
Theta Ophiuchids	May 21–June 16	June 10	10
June Lyrids	June 10–June 21	June 15	up to 8
Pi Sagittariids	June 1–July 15	June 18	5
Ophiuchids	May 19–July 2	June 20	6
June Bootids	June 27–July 5	June 28	1–2
June Scutids	June 2–July 29	June 27	about 2–4

*Major showers shown in bold.

Meteor Shower	Active Period	Peak	ZHR Max
Alpha Lyrids	July 9–July 20	July 14	1–2
Southern Delta Aquarids	July 14–August 18	July 29	15–20
Alpha Capricornids	July 15–September 11	August 1	6–14
Southern Iota Aquarids	July 1–September 18	August 6	7–8
Upsilon Pegasids		August 12	1–2
Perseids	July 17–August 24	August 12/13	50–60 minimum
Northern Delta Aquarids	July 16–September 10	August 13	up to 10
Alpha Ursid Majorids	August 9–August 30	August 13–14	4
Kappa Cygnids	July 26–September 1	August 18	up to 6
Northern Iota Aquarids	August 11–September 10	August 25	5–10
Alpha Aurigids	August 25–September 6	September 1	up to 9
Gamma Aquarids	September 1–September 14	September 7	1–4
Southern Piscids (Piscids)	August 12–October 7	September 11	up to 5
Eta Cetids	September 20–November 2	First week of October	
Delta Aurigids	September 22–October 23	October 6–15	
Autumn Arietids	September 7–October 27	October 8	3–5
Draconids (Giacobinids)	October 6–October 9	October 8	1–storm
Epsilon Geminids	October 10–October 27	October 19	1–2
Orionids	October 15–October 29	October 21	25–30

Meteor Shower	Active Period	Peak	ZHR Max
Southern Taurids	September 17–November 27	October 30–November 7	7 (Northern and Southern Taurids combined)
Northern Taurids	October 12–December 2	November 4–7	
Andromedids (Bielids)	September 25–December 5	Around November 14	up to 5
Leonids	November 14–November 21	November 17	10–15+
Alpha Monocerotids	November 13–December 2	November 21	1–5
Delta Arietids	December 8–January 2	December 8–9	1
Chi Orionids	November 16–December 18	December 10	5
December Monocerotids	November 9–December 18	December 11	1–2
Geminids	December 6–December 19	December 13–14	50–100
Coma Berenicids	December 12–January 23	December 19–29?	
Ursids	December 17–December 25	December 22–23	10–20
Zeta Aurigids	December 11–January 21	December 31	

4

Meteorites

THUNDERSTONES

Meteorites have been recognized and collected for millennia. Meteoritic iron was used to make tools as early as 4,000 B.C. The Egyptians referred to these "heavenly irons" on the interior walls of several of the pyramids. Greek philosophers, including Anaxagoras and Plutarch, referred to meteorites in their writings, and Aristotle incorrectly stated that meteors and fireballs were weather phenomena. The Romans Livy and Pliny also mentioned meteorites in their writings.

In Mecca, Muslim pilgrims will find the Hadshar al Aswad or Black Stone. Considered by some to be a meteorite, legend has it that Mohammed wept when he touched the stone.

Meteorites have been recorded in many other civilizations, including the Japanese (as religious artifacts), Chinese (recorded meteorite falls), and the Europeans, including the famous 1492 A.D. Ensisheim fall. In the United States alone, at least twenty-two verified building strikes were recorded in the twentieth century.

In the seventeenth and eighteenth centuries, people referred to meteorites as thunderstones. They thought these rocks were volcanic in nature and had simply been struck by lightning!

METEORITE ORIGINS

Several sources of meteorites are found in the solar system: asteroids, comets, and meteoroids, as well as dust and other material left behind by comets. Any of these materials can enter the earth's atmosphere, but the question is, will the object survive the fiery descent and make it to the earth's surface?

The recently-observed impact of Comet Shoemaker-Levy 9 fragments into Jupiter demonstrates that cometary impacts still occur in our solar system. In fact, there is evidence that such impacts have occurred on a fairly regular basis throughout the solar system.

The most recent possible major cometary impact on the earth was most likely in Tunguska, Siberia, on June 30, 1908. (One other currently accepted theory regarding the Tunguska impact is that it was a large but fragile stony meteorite.) The explosion was heard as far as six hundred miles away. Trees were knocked down in a radial pattern for a distance of twenty miles from the center of impact. It was years after the Tunguska impact that a research team finally made it to the site, but no significant material was recovered from the impact.

One of the most famous meteorites that is believed to be of cometary origin is Orgueil, which fell in Montauban, Tam-et-Garornne, France, on May 14, 1864. Orgueil fragments look like pieces of charcoal briquettes used for barbecuing—and are very expensive!

Considering the earth's water-land ratio, we see that the

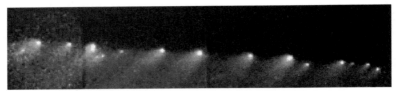

Hubble Space Telescope image of Comet Shoemaker-Levy 9 before impact—the "String of Pearls." COURTESY OF THE SPACE TELESCOPE SCIENCE INSTITUTE, NASA

chances of a meteorite falling on land are only about one in four. And many land areas are difficult to get to and explore, such as dense rain forests, high mountains, trackless deserts, and the like. Most of us do not have everyday access to the earth's polar regions (though a significant number of meteorites have been recovered in Antarctica). Meteorites can become buried, not only from impact, but also from blowing and shifting sand, agriculture, and the like. Finally, erosion can get the best of a meteorite, making it almost indistinguishable from terrestrial rocks.

Comet Shoemaker-Levy 9 Jupiter impact sites taken by the Hubble Space Telescope planetary camera, July 1994. Eight impact sites are visible. COURTESY OF THE HUBBLE SPACE TELESCOPE COMET TEAM, NASA

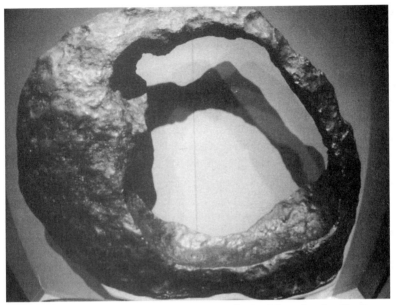

The Tucson Ring meteorite, on display at the Smithsonian Institution National Museum of Natural History. The Tucson Ring, an Iron (UNGR) Ataxite, Ni-rich, is sometimes referred to as the Signet or Irwin-Ainsa Iron. PHOTO BY DR. MIKE REYNOLDS

This is not to dissuade you from looking for meteorites or joining the ranks of the meteorite collector. It is meant to point out the rarity of these extraterrestrials. Those of us who have found a meteorite on the ground, purchased a pristine, fresh sample, or felt the smooth surface of a just-polished and etched meteorite will tell you that it is an exciting, but challenging endeavor. The best and most successful collector in any field knows as much as possible about the items he or she is collecting.

FALLS AND FINDS

If you are fortunate enough to see a bright meteor or fireball fall to earth and then find the impact site, you have come across what is referred to as a fall. Often the meteor or fireball is observed but the point of impact is not. If many people saw the fireball, triangulation can be used to determine the point of impact. Then a search team will go out and look for the meteorite. Careful searching might—or might not—turn it up. Often it takes several trips to the suspected fall site, and frequently the meteorite is never found. Occasionally people stumble on the meteorite, and sometimes in a most unusual way. (Meteorites have been known to impact homes, cars, mailboxes, roads, dogs, and even people.)

An historic meteorite fall occurred in 1492 in the village Ensisheim. The fall of the meteorite was signaled by a loud explosion. History records that a boy actually saw a single stone fall from the sky and land in a wheat field. People quickly surrounded the meteorite, which weighed around 150 kilograms. Eventually the meteorite was lifted out of a hole one-yard deep and placed in front of a church in the town. People began breaking off pieces of the meteorite for good luck (history notes that the mayor put a stop to that).

King Maximilian of Austria heard of the Ensisheim fall, and since he was engaged in a battle nearby with the French, the King decided to take a look at the meteorite. Not wanting to miss a good opportunity, the King decided that the fall was a sign from God, foretelling of his upcoming victories in his battles with the French. After removing a piece for himself, Maximilian declared that the Ensisheim meteorite should forever be kept at the church. King Maximilian and his army went on to defeat the French at the battle of Salins.

The Ensisheim meteorite remained in the church until the time of the French Revolution, when it was taken to a museum in nearby Colmar by French revolutionaries. As Ensisheim citizens did many years earlier, French scientists removed some samples for study. The meteorite was eventually returned to the church but by this time had lost considerable mass. The stone was eventually moved to the Ensisheim Town Hall, where it remains to this day.

The largest recorded meteorite fall occurred near Pultusk, Poland, on January 30, 1868. A bright fireball was first seen, which exploded loudly. Meteorites literally showered the villages to the east of Pultusk. It has been estimated that more than one hundred thousand meteorites fell. Many of them were only a few grams in weight, leading to the term "Pultusk peas."

A major fall was witnessed on February 12, 1947, in the Sikhote-Alin Mountains, Siberia, when thousands of iron meteorites fell, some forming craters as large as football fields. The resulting shock wave was felt one hundred miles away. Over twenty-five tons of shrapnel-like meteorites have been recorded.

On December 10, 1984, thirty-six meteorites fell in Claxton, Georgia. One of the meteorites, a 1,455-gram specimen, struck a mailbox and knocked it to the ground. Talk about special delivery! Not only did the meteorites become collectibles, so did the mailbox.

Peekskill New York's famous meteorite chose to impact a Chevrolet Malibu. On October 10, 1992, a 12,370-gram meteorite struck the parked car of Michelle Knapp. Initially the police thought the car had been bombed, or that a piece of an airplane

The special delivery damage caused by a meteorite in Claxton, Georgia.
PHOTO BY BLAINE REED

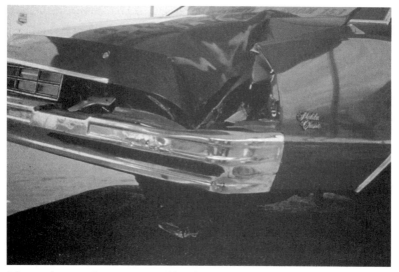

The car impact damage caused by the October 10, 1992 meteorite fall in Peekskill, New York. PHOTO BY MIKE MARTINEZ

had fallen out of the sky. The car toured the world and has become as much of a collectible as the meteorite. Before impact, thousands of people gathering to see high school football games saw a bright fireball. Videos of the fireball showed its trajectory: from the asteroid belt all the way to Michelle Knapp's Malibu!

The police didn't have far to go when investigating the fall of three meteorites in Northern Ireland on April 25, 1969. Two pieces weighing a total of 513 grams struck the Sprucefield police station. The third meteorite fell in a field near Bovedy.

On February 8, 1969, in the area surrounding Chihuahua, Mexico, a very bright fireball streaked across the sky just after 1:00 A.M. This fireball, brighter than the full moon, literally showered the little city of Pueblito de Allende, south of Chihuahua, with thousands of meteorites ranging from pebble- to football-size. The Allende fall produced over two tons of meteorites. The local Indians of Mexico's high country consider these extraterrestrials good luck charms from the sky.

A question often asked is, "How many people have been hit or killed by meteorites?" Surprisingly, a few people have been hit, and no one has been killed, even though it has been rumored throughout history that there may be some who met a cosmic end.

Probably the most famous recent human strike was that of Mrs. E. Hewlett Hodge of Sylacauga, Alabama. On November 30, 1954, Mrs. Hodges was relaxing on her couch at home, across the street from the Comet Drive-in Theater, when a 3.9-kilogram meteorite came through her roof, ricocheted off a radio, and struck her left thigh before coming to rest in her living room. Mrs. Hodges could not even claim the meteorite as her own, for the home was rented. The landlord took the meteorite, and all that Mrs. Hodges received was a deep bruise and a rather embarrassing photo on the cover of *Life* magazine.

The town of Weathersfield, Connecticut, had two meteorite impacts over an eleven-year period. On April 8, 1971, a 286-gram meteorite crashed through the roof of a home while the residents were asleep. The next morning they found a black rock sticking through their living room ceiling.

The second Weathersfield impact occurred on November 8, 1982, when a meteorite struck a different house across town. This meteorite went through the roof into the living room, bouncing into the dining room and landing under a table—all while the residents were watching television.

One of the most unusual falls occurred on June 28, 1911, in Nakhla, Egypt. About forty meteorites fell with a total weight of approximately forty kilograms. One of these meteorites purportedly struck and killed a dog. The Nakhla meteorites are most desirable because they are believed to be of Martian origin, and thus are classified as a Mars rock. A small vial with Mars rock *dust* can sell for $100.

Most meteorites are discovered as a find. They are not seen to fall but are simply discovered some time afterward. The ratio of falls to finds is about 1:2. People will often think unusual-looking and/or heavy rocks, especially slag (the waste material remaining after metal has been melted down from ore), are meteorites. Many meteorite dealers, and those who spend time in the field collecting samples will tell you of all the objects sent to them claimed by their owner to be meteorites, which were, in fact, not.

Finds often result from farming activities. Plowing turns up new soil, and farmers prefer to move rocks—even meteorites—out of their fields. Meteorites are easier to find in some areas than in others; due to the surrounding geology, or lack of competing rocks, the meteorites simply stand out.

Probably one of the most successful meteorite hunters was the late Harvey H. Nininger. A biologist and college professor at a small Kansas college, Nininger first began collecting meteorites in 1923. He would go to an area, educate the local residents as to what a meteorite looks like, and offer them a finder's fee. One of Nininger's early expeditions to locate and buy meteorites took him to Xiquipilco, Mexico, a known meteorite fall area. While enjoying breakfast with El Presidente, the mayor of the little village, Nininger asked about *meteorito,* the Mexican word for meteorite. The mayor left and returned with a beautiful nine-kilogram specimen. Nininger, who could barely contain his excitement, asked if he could get more meteorites like the nine-kilogram beauty. The mayor replied that he could probably get Nininger a ton of these meteorites.

Robert Haag has traveled worldwide collecting meteorites—and has stories to go with his many adventures. Called the Meteorite Man by many, Haag has endured airplane crashes, an arrest, paragliding, and a lot of horse trading in search of meteorites. One of the easiest meteorite recoveries made by Haag was in La Criolla, Argentina. The local constable not only assisted, but wanted to participate in the search. Their stops included the mayor's home, where Haag purchased a beautiful, 6,100-gram stony meteorite, a local coffee shop, and a farm (7,000- and 2,600-gram specimens). By the end of the first trip, Haag had recovered fifty-four meteorites with a combined weight of 34 kilograms. Why were the residents of La Criolla so willing to part with their space rocks? Partly, they were terrified of the meteorites! In addition, Haag made each owner a good offer.

Meteorite dealer Blaine Reed tells the story of his search for meteorites at the side of the road. In areas that have few natural rocks, it makes sense that any rock found might be a meteorite. The local farmers would collect rocks as they plowed, then dump them at the ends of their fields, often in a ditch at the edge of a road. Reed thought it might be a good plan to drive along the road and scan the farmers' deposits—until he nearly had a serious automobile accident.

Reed had better luck in Texas, searching a farmer's field with a magnetometer. His search took him under high tension electrical wires, at which time his detector started "screaming." Reed, thinking he had made a big find, began digging and dug several feet with no find. The farmer finally stopped by and told Reed that the area had just been struck by lightning—the soil had been magnetized. The lightning strike had been directed by the overhead electrical lines.

Not to be dissuaded, Reed changed direction and kept searching. A short while later his persistence paid off as the detector began to scream again. After some digging Reed found a large meteorite.

Dealers sometimes stumble on meteorites being used as interesting household items. A 60-kilogram meteorite found in the home of a local rancher in Guadalupe y Calvo, Mexico, had been used for over twenty years as a dog food bowl! After the man sold the meteorite to a dealer, he got a new pickup truck and the dog got a new bowl.

The Gan Gan meteorite was also an unusual find. A man in Argentina collecting pine cones first found an 83-kilogram meteorite in 1984. However, it wasn't recognized as a meteorite until 1998, 14 years later.

I recall an incident I investigated where a Southern Georgia farmer claimed to have seen a meteorite fall and actually split a tree in half. After a long drive to rural southern Georgia, I determined that the "meteorite" was a large, ant-infested piece of iron slag. The farmer was so upset that he threatened bodily harm and legal action, thinking my intention was to dupe him out of his lottery-win retirement rock. The farmer later contacted a "real meteorite dealer," who informed him that it was indeed slag.

What are the tools of the collector? First of all, some general information about the location of a find, or fireball/bolide data that indicates the general area of a fall. A friendly demeanor toward the locals is often helpful in gleaning information, especially in the case of foreign finds and falls. People who live in or near the area of the find/fall may be able to take you right to the spot; they might also be able to supply samples—often at a price.

Many collectors use specific tools in the field: a good-quality metal detector, a shovel or pick, a GPS device to determine the exact location of the find, a camera to record the find, tongs to pick up the meteorite, (some do not want to touch it until lab verification) and a magnet. If using a metal detector, keep it set at maximum intensity. Dress accordingly for the area and remember to protect yourself against the elements.

You should always determine if it is legal to collect meteorites in the area where you are searching. For example, in the United States any meteorite found on public property belongs to the federal government and must be sent to the Smithsonian Institution's National Museum of Natural History. Owners of private property can legally deny you access; many people have heard about the value of meteorites and, like the case of the Southern Georgia farmer, think they have found their fortune. Some countries have barred any meteorite collecting or exporting whatsoever. And the meteorite you find just might be a local or national treasure—ask Robert Haag.

As the story goes, Haag had heard about a meteorite in Chaco Province, Argentina, that was available for purchase. The problem with this meteorite was its weight: 37 tons. So Haag made arrangements for purchase, flew to Argentina to meet the seller and "pick up" his meteorite for transportation. However, Haag found out that the dealer was trying to sell him a local treasure! Luckily Haag was able to clear himself, but sadly lost a big meteorite acquisition.

The best tool is probably a good eye. Be able to recognize meteorites, and know their characteristics (see chapter 5). Study meteorites in your local museum, those of other collectors, and meteorites you purchase. Keep meteorwrongs—pieces of slag, nodules, and the like that might look somewhat like meteorites but are not (thus the term meteorwrongs, commonly used by meteorite collectors)—so you can compare your "find" to other samples. As a new collector, knowledge of meteorwrong characteristics will not only help you as you search for meteorites, it will help you when you buy specimens for your collection.

5

Meteorite Classification

TYPES OF METEORITES

Meteorites can be classified into three basic categories: irons, stones, and stony-irons. One very important fact to keep in mind is that all meteorites are magnetic. Those who work in the field looking for meteorites keep a magnet with them at all times. This is always the first test; if the meteorite suspect is not magnetic, it is a "meteorwrong." Good collectors learn very quickly how to identify meteorwrongs since the purchase of what may be represented as a very rare stone meteorite (at $200 a gram!) could turn out to be an expensive piece of junk.

The classic meteorwrong is an igneous rock known as a Cumberlandite. For years, Cumberlandites have been found in Rhode Island, originating from an outcrop in Cumberland, Iron Mine Hill. Cumberlandites are still being offered for sale as meteorites.

Meteorite specimens are almost always recorded by weight using the metric system. You will find weights in kilograms (kg) or grams (g). Recalling your English to metric system conversions, there are 454 grams in a pound and about 2.2 pounds in a kilogram.

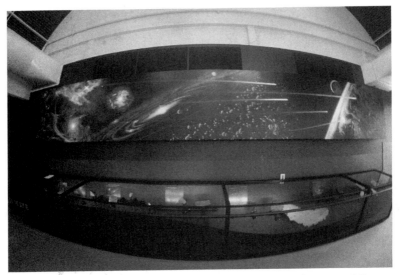

The "Meaning of Meteorites" mural by artists B. E. Johnson and Joy Day accents and explains the meteorites in the exhibit at the Chabot Space & Science Center. PHOTO BY DAVID SISSON

There are four visual clues in recognizing a meteorite:
1. It attracts a magnet (contains metallic iron).
2. It has a usually dark, thin, melted surface layer (called a *fusion crust,* due to the melting of minerals on the meteorite's surface during the plunge through the earth's atmosphere).
3. It has an aerodynamic shape (acquired during the high speed flight).
4. It has thumbprint-shaped indentations (called *regma-glypts,* made during the entry through the atmosphere).

Iron meteorites are the easiest type to identify because of their weight and metallic nature. The most durable of all meteorites, they are the most likely to be collected first. Many collectors got their start as children when they bought a small piece of an iron meteorite from a museum gift shop. However, iron meteorites make up only about ten percent of all meteorite falls.

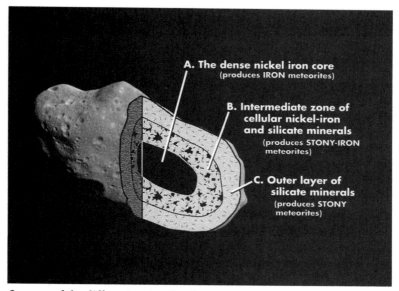

Sources of the different meteorite classes. ILLUSTRATION BY DAVID FRANTZ

Iron Meteorites

Iron meteorites, or irons, as they are commonly known, are actually a combination of iron (Fe, the chemical symbol for iron), nickel (Ni), and cobalt (Co). Iron is the predominant metal, combined with five percent up to fifty percent nickel (as an alloy with iron in most iron meteorites), and a trace of cobalt. Iron meteorites are divided into three basic groups, depending on the ratio of iron to nickel and the rate at which the parent body originally cooled. (In actuality, there are twelve classes of irons and over fifty groups of irons that belong to specific asteroids.)

Hexahedrites

The word "hexahedrite" refers to a type of crystalline form, a hexahedron or six-sided crystal. Hexahedrites contain between 4.5 and 6.5 percent nickel and large crystals of kamacite, an iron-nickel crystal containing up to 7.5 percent nickel. Other elements are often also found in minute quantities, such as carbon (C),

The author with the Goose Lake meteorite, an Iron Medium Octahedrite (IAB), Smithsonian Institution's National Museum of Natural History. Earle Lindsley, Director of Chabot Observatory, recovered the meteorite in 1937. PHOTO BY JOSE OLIVAREZ

chromium (Cr), cobalt (Co), phosphorous (P), silicon (Si), and sulfur (S).

Hexahedrites produce a series of lines or bands called *Neumann lines*, first detailed by Johann Neumann in 1848. These are a very fine series of lines, sometimes crossing each other. Neumann lines are seen when the sample is cut, polished, and acid etched. They are the result of high pressures and temperatures.

Octahedrites

Octahedrites contain octahedral, or eight-sided, crystals. In addition to the iron and nickel matrix, octahedrites also can contain chromite, cohenite, diamond, and schreibersite, along with nodules of graphite and troilite, which are often surrounded by kamacite.

Octahedrites produce a wonderful crystalline pattern of lines after being cut, polished, and etched with a weak solution of nitric acid (HNO_3). This pattern is called the *Widmannstatten*

Gibeon meteorite, machined into an 882-sphere and etched, showing the Widmannstatten pattern. Note the magnet—a tool most meteorite collectors keep nearby. PHOTO BY DAVID SISSON

The Willamette meteorite, an Iron Medium Octahedrite (IIIAB), at the American Museum of Natural History. COURTESY DEPT. OF LIBRARY SERVICES, AMERICAN MUSEUM OF NATURAL HISTORY, PHOTO BY P. HOLLEMBEAK, NEG. NO. 3316

pattern or Widmannstatten figure after its discoverer, Count Alois von Widmannstätten, a porcelain manufacturer from Vienna, who described the pattern in 1808. (Actually William Thompson first discovered the figures in 1804, but the pattern was named after Widmannstätten.) The Widmannstatten pattern is produced by broad kamacite bands sandwiched between narrow taenite bands, which are parallel to the octahedron's faces. During the acid etching procedure, the kamacite bands, which are nickel poor, are more altered by the acid than the nickel-rich taenite bands. This produces a somewhat relief-like look (a three-dimensional effect). The Widmannstatten pattern is not found in any earth rocks nor can it be replicated in the laboratory. It is unique to iron meteorites.

The grouping of Octahedrites, from finest to coarsest, is determined by the width of the band or line in the Octahedrite's Widmannstatten pattern.

The coarsest octahedrites are considered "rusters." Collectors can see these specimens literally fall apart into a pile of rust while sitting in their collection. This unfortunate event can occur

TYPES OF OCTAHEDRITES

Octahedrites are divided into several major groups, depending on the iron to nickel ratio and the kamacite-rich band width (Widmannstatten Pattern).

Octahedrite Type	Width of Band	Percent Nickel	Availability
Coarsest	greater than 3.3 mm	6.5 to 7.2	rare
Coarse	1.3 to 3.3 mm	6.5 to 7.2	common
Medium	0.5 to 1.3 mm	7.4 to 10.3	most common
Fine	0.2 to 0.5 mm	7.8 to 12.7	common
Finest	less than 0.2 mm	7.8 to 12.7	rare
Plessitic	less than 0.2 mm	Kamacite twists or spindles	rare

Sikhote-Alin, an Iron Coarsest Octahedrite (IIAB) meteorite that impacted in Russia in 1947. PHOTO BY DAVID SISSON

An Odessa, Texas, Iron Coarse Octahedrite (IAB) meteorite slice that shows rust, in spite of all attempts to protect the sample. PHOTO BY DAVID SISSON

even in low-humidity environments, so proper specimen storage is paramount.

Ataxites
Ataxites are also called silicated irons. This class of irons contains significant amounts of nickel—greater than sixteen percent—forming a unique nickel-iron alloy. Ataxites do not exhibit Neumann lines or a Widmannstatten pattern when cut, polished, and etched with nitric acid.

Stone Meteorites
Stone meteorites, also called stony meteorites or stones, make up over ninety-four percent of observed falls. Most are thought to be material from the crust and mantle of asteroids; a few stony meteorites are thought to be from comets. Stone meteorites contain approximately seventy-five percent to ninety percent silicate materials, such as olivine and pyroxene, which contain silicon, oxygen, and one or more metals. Most stones also contain an iron-nickel alloy. There are two major groups of stony meteorites: chondrites and achondrites.

Chondrites
Chondrites are so named because they contain millimeter-size spherical crystals of minerals such as olivine and pyroxene, called *chondrules,* imbedded in the stony material. Edward Howard and Jacques-Louis Comte de Bournon first described chondrules in 1802. The term comes from the Greek *chondros,* meaning a grain of seed.

Chondrites can be as porous as sandstone, and easy to break or crush. There are several subgroups:

- *Amphoteries,* or LL chondrites, contain very little iron ("LL" stands for low iron and low metal); they tend to be composed of fragmented rock.
- *Carbonaceous,* or C chondrites, contain organic compounds. Carbonaceous chondrites are very rare, contain little metal, exhibit well-defined chondrules, and tend to have a black

to gray matrix. There are four subclasses of carbonaceous chondrites (C1 through C4) depending on their composition and state of alteration.

- *Enstatite,* or E chondrites, contain the silicate enstatite, an iron-free pyroxene. Two subclasses (H and L) of enstatite chondrites are dependent on iron content. Enstatite chondrites are quite rare.
- *Olivine-bronzite,* or H chondrites, the most abundant class of meteorites, contain a high degree of iron (thus the "H") both in metal flakes and mineral form.
- *Olivine-hypersthene,* or L chondrites, contain less iron than olivine-bronzite chondrites.

Some additional classes of chondrites have been suggested, including B chondrites (more than fifty percent iron-nickel alloy) and R chondrites (highly oxidized, olivine-rich, and little iron-nickel alloy).

Some of the stones from the Juanchenge, China, fall in 1997. Juanchenge is an Ordinary Stone Chondrite (H5). PHOTO BY DR. MIKE REYNOLDS

Two samples of Millbillillie, an Australian Stone Eucrite (EUC). Note the lines on the right-hand sample, indicating the ground level where the meteorite was buried. PHOTO BY DR. MIKE REYNOLDS

Achondrites

Achondrite meteorites lack chondrules or metal flakes, but are rich in silicates. They are considered to be very rare. Achondrites are believed to have undergone advanced geological processing on their parent bodies, such as lava or magma flows and impact breccias. There are many subgroups:

- *Eucrites*, the most abundant of the achondrites, are calcium-rich basaltic meteorites, meaning they are like volcanic rocks. Eucrites are nevertheless chemically different from terrestrial basalt and appear to be from the asteroid Vesta.
- *Diogenites* (calcium-poor basaltic meteorites) are related to the eucrites. They also appear to be from the asteroid Vesta.
- *Aubrites* are calcium- and iron-poor meteorites consisting mostly of enstatite (aubrite might be related to the enstatite chondrites).

- *Ureilites* are calcium-poor meteorites consisting mainly of olivine, pyroxene, and carbon in the form of either graphite, diamond, or lonsdaleite (a rare pure carbon mineral like diamond, but with a different crystal structure), as well as a significant amount of iron and nickel.
- *Lunar meteorites* have been found in small numbers. All except two, found in Australia and the Libyan Sahara desert, were recovered in Antarctica. Lunar meteorites are impact breccias—rocks formed by the rewelding of loose fragments that were shattered during impact events. Lunar meteorites may be identified by a fusion crust with slightly green hues and by a gray interior with angular inclusions of often brighter materials. Lunar meteorites are believed to have been blasted off the moon as ejecta from high-velocity impact events.

Mars rocks—on earth! Left: a 3.6-g Zagami from Nigeria. Right: a 2.128-g Dar al Gani 476 from Libya. These two specimens, both Stony Achondrite Shergottites (SNC), together are valued at well over $2,000.

PHOTO BY DAVID SISSON

The now-famous Antarctic Mars rock, ALH84001. PHOTO COURTESY OF JOHNSON SPACE CENTER, NASA

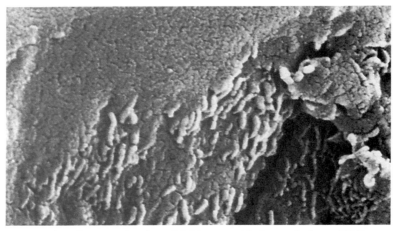

The tubelike form on the above highly magnified sample from ALH 84001 was suggested by some to be a fossil of a simple Martian organism that lived over 3.6 billion years ago. PHOTO COURTESY OF JOHNSON SPACE CENTER, NASA

- A few apparently Martian meteorites have been found. These achondrite Mars rocks, commonly referred to as the *SNC subgroup* (after Shergotty, Nakhla, and Chassigny) are distinguished on the basis of mineralogy, but all share characteristics that together point to a Martian origin.

A number of achondrites do not fit into any of the preceding groups or subgroups. Some are the only meteorite known of their kind. One of these is the Antarctic meteorite ALH84001. Identified as a Martian meteorite, ALH84001 was made famous by the 1996 announcement that carbonate minerals and globules contained within it had formed under the influence of Martian microorganisms. In December 2000, a new report described tiny magnetite crystals in ALH84001, identical to those used by aqueous Earth bacteria as compasses to find food and energy. However, not all scientists are convinced and have placed doubt on the microorganism theory.

Stony-Iron Meteorites

Stony-iron meteorites, or stony-irons, are composed of silicates and iron-nickel alloy in roughly equal proportions. Only about one percent of all falls have been stony-irons. There are three main groups: pallasites, mesosiderites, and lodranites.

Pallasites

Pallasites consist of olivine crystals embedded in an iron-nickel alloy matrix. The olivine crystals can be as large as three-eighths inch across. Cut and polished pallasite sections are simply the most beautiful form of meteorite representation. Pallasites are believed to have been formed at their parent body's core-mantle boundary. The iron-nickel matrix (octahedrite in nature) would have come from the core and the olivine from the mantle's base. Pallasites are the most common of the stony-iron meteorites.

Mesosiderites

Exhibiting characteristics of multiple impacts, mesosiderites, or mesos, are often called the "wastebasket" or "dumping ground" form of meteorites. The iron-nickel alloy of mesos does not form

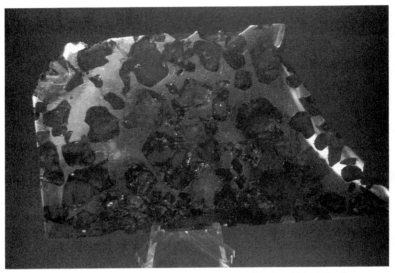

A 138–gram slice of the famous Stony-Iron Pallasite (PAL) Imilac meteorite from the Atacoma Desert, Chile. PHOTO BY DR. MIKE REYNOLDS

A slice of a Vaca Muerta Stony-Iron Mesosiderite (MES), Taltal, Atacoma, Chile. Note the chondrule on the slice's lower right. PHOTO BY DAVID SISSON

a matrix as in the pallasites, but rather is limited to grains and nodules. Like the pallasites, the iron-nickel alloy exhibits the same properties as octahedrites.

Lodranites

A very rare meteorite type, lodranites consist of roughly equal amounts of olivine, pyroxene, and iron-nickel alloy. Lodranites are coarse and easily broken.

6

Collecting Meteorites

Why collect meteorites? Collecting is as old as the human race. Today, people collect everything from baseball cards to coins, stamps, and Beanie Babies; you name it and someone probably is collecting it.

An article from the February 1, 1999 issue of *U.S. News & World Report* ("Forget Beanie Babies, These Rocks Are Hot!") noted that meteorites have become the next collectible. Meteorites—especially those that are known to have come from the moon and Mars—are bringing in record proceeds, even at auction houses. The article notes that "the craze to own space rocks and resulting price inflation aren't likely to end soon."

Like so many collectibles, there is a limited supply of meteorites, and the "supply" is very unpredictable! The stories behind meteorite falls are often as interesting as the geology and chemistry of the meteorite itself.

Meteorites as a collectible also bring their risks, and the collector really needs to know the field or really trust his or her meteorite dealer. If a dealer states a particular meteorite is a "Mars rock," can you, as a collector, verify the specimen upon inspection? Even the best of meteorite dealers can buy what turns out to be a poor meteorite for a collection or as an investment. A perfect example of this was the distribution and sales of the Chinese meteorite Nantan. Collectors and dealers were delighted to have Nantan meteorites in their collections or as specimens for sale, until it was discovered that a Nantan meteorite is a terrible "ruster"; that is, it literally rusts "to dust" while sitting on your shelf!

Even with the sometimes inflated prices and the risks of purchasing a "meteorwrong," there is something special about holding—and owning—a rock that fell from outer space. After all, how many people own a piece of the planet Mars? Meteorites can be interesting in shape, color, and exterior appearance, or when cut or polished. Among the billions of rocks on the earth's surface, there is a limited number among them of fairly extraordinary composition and that arrived here in an unusual way. Today we can own, literally, a piece of the *space* rock for ourselves. Indeed, the uniqueness of a meteorite collection and each specimen's beauty and story can be inspiring. The preservation of meteorites is important, both in the hands of the private collector and in museums and universities.

The Hoba meteorite, the largest intact meteorite yet discovered on the earth, near Grootfontein, Namibia, is prominently featured on a postage stamp. PHOTO BY DAVID SISSON

PURCHASING METEORITES

It would be nice to start a collection by finding a meteorite your-self! Unfortunately, that is unlikely for the majority of new col-lectors, so we end up buying a meteorite or two from dealers. Perhaps you were one of the few who bought a small meteorite at a local museum as a child. What a fantastic acquisition that must have been for you—and what a great way to start your collection!

You should initially consider buying small examples of dif-ferent types of meteorites for your collection. That is, try to start your collection with a nice iron, a nice stone, and a nice stony-iron. A few collectors specialize in collecting one particular type of meteorite. But the majority of collectors have representative examples of several types of meteorites in their collections.

Meteorites are usually sold by the gram or kilogram weight. Don't be too shocked at the high prices per gram at which some meteorites are sold. Appendix F, which lists thirty meteorites and information about each one (including price per gram as of late 2000), will provide you with a starting guide.

Meteorites can be purchased whole or as a sample slice. Whole samples of irons are fairly common, and an iron slice (one that has been cut, polished, and acid etched to show the Wid-mannstatten pattern or Neumann lines) is a beautiful and inter-esting piece to own.

Stony and stony-iron meteorites are more often available for purchase as cut and polished slabs. These slabs show the mete-orite's internal structure including chondrules and other inclu-sions. Some collectors prefer an end piece because it shows the internal as well as the external structure, such as the fusion crust.

A whole stony meteorite adds to the collection's diversity. Small, affordable ones are still available on the market.

The true beauty of a stony-iron is seen in a slice that has been polished. Lit from the back, the olivine crystals within this type of stony-iron meteorite (surrounded by the shiny iron-nickel matrix) are simply beautiful.

When buying a meteorite, consider the quality of the sample as well as the price. What can you afford? You might initially buy a Sikhote-Alin or Gibeon iron meteorite sample because you cannot afford a Nakhla sample (Nakhla meteorites are at least five hundred times more expensive per gram than Sikhote-Alin or Gibeon meteorites). You can usually find beautiful examples of various types and sizes, so find something that appeals to you. Many collectors collect based on what they find personally appealing.

Meteorites are available in all sizes and weights and from mediocre to fine quality. A large, museum-quality piece will cost you thousands, if not tens or hundreds of thousands of dollars. Rare or unusual falls and finds will cost even more. Collectors also show great interest in objects actually struck by meteorites: cars, mailboxes, and even a section of a road! The interest lies in the fact that these impacted objects also tell a story.

You should get to know at least one or two meteorite dealers that you can trust to assist you in making purchases. There are many meteorite dealers, both in the United States and abroad. A listing of some of the U.S. dealers appears in Appendix D. It does pay to shop around; prices can vary significantly from dealer to dealer.

Meteorites are also available for purchase on the Internet through on-line bidding. You can get good deals on excellent samples this way, although you will not be able to inspect the meteorite in person. A seller will often post an image of the meteorite that is up for bid; knowing the seller—is he or she a reputable seller—will help you to determine if the meteorite is "as advertised." A number of bogus items have come up for bid, so the on-line bidder needs to know the field.

You will find the prices of some irons to be very reasonable. Stony meteorites can also be purchased at a reasonable price, although they are generally (gram for gram) more expensive than irons. And there are some rare stony meteorites that cost thousands of dollars per gram. On the other hand, a nice stony-iron meteorite slice will cost you hundreds of dollars but is well worth the investment.

An affordable way to collect meteorites is by purchasing

micrometeorite or bug box samples. These are small specimens that display the meteorite's characteristics. Micrometeorites are usually stored in small plastic boxes, or bug boxes, some of which feature a magnifying lens cover.

PRESERVING YOUR METEORITES

Once you make your purchase, you will want to protect the sample from the elements. Humidity, oxygen, and pollution in the atmosphere damage meteorites. It is beneficial if the meteorite was quickly recovered after the fall; weathering can destroy a meteorite.

Some collectors, and especially museum and university lab personnel, always use disposable gloves when handling meteorites. Moisture and oils from the hands can damage a specimen in the long term. So preserve your meteorites by not overhandling them with your bare hands.

A freshly recovered meteorite (should you be lucky enough to find one) should be carefully cleaned. It is recommended that a museum, university, or meteorite dealer clean the specimen; novices can unintentionally damage the meteorite while attempting to clean it. Meteorites purchased from reputable dealers will already be cleaned to the extent that cleaning it is possible.

Rust is a serious problem, and one that can develop years after acquiring an iron meteorite. Look for discoloration, usually brown, red, or yellow in appearance. Rust is especially a problem with coarse octahedrite irons.

If rust is discovered early enough, it is possible to remove much or all of it by using a diluted acid solution. Use one part acid to five or six parts water; oxalic acid, sodium citrate, sodium hydrosulfate, or tartaric acid can be used for this procedure. Soak a cotton ball in a small quantity of the acid solution and apply the cotton ball directly to the rust for ten to twenty minutes. After the acid application, soak the meteorite in an anhydrous alcohol bath for a couple of hours. Finally, apply a clear lacquer coat as a sealer (multiple thin coats without runs works best). If you need help, ask a reputable meteorite dealer to assist you. In any case, keep water in any form away from your meteorites.

A meteorite left in the open should periodically be carefully dusted. A soft photo lens brush or camel hair brush will do the job. A lens squeeze bellows can also be used to remove dust; blowing on the sample will deposit moisture from your breath on your meteorite. Both a lens squeeze bellows and a photo lens brush can be purchased from a local camera dealer.

STORAGE AND DISPLAY

There are a number of ways to display and store your meteorites:

- *Bug boxes.* As discussed earlier, bug boxes are an excellent way to display and store small specimens. (See Appendix D for a supplier.)
- *Caliper stands.* These exhibition stands are popular and attractive, and even available in 24-kt. gold-plated brass. The entire meteorite can be viewed on these stands except where the meteorite is held by the calipers. The only problem with this display method is dust.
- *Custom boxes.* A set of same-size custom-made boxes, with a plastic face, are an attractive and safe way to store your meteorites. In these boxes, the meteorites can be handled or passed around with safety. However, unless you are a skilled woodworker or know someone who will make custom boxes for you, this could be an expensive proposition. Riker specimen mounts (discussed below) are an excellent alternative.
- *Lucite.* Kits for totally encasing specimens are available at local hobby shops. Lucite kits include a resin and hardener. Molds are also available, as well as coloring agents. Although providing the best protection, the meteorite cannot easily be removed later without damage; therefore many collectors stay away from Lucite.
- *Peg stands.* Plastic stands, much like small easels or rings, are often used as an attractive and inexpensive way to display specimens. Like caliper stands, the main problem is dust. Also there is more of a tendency to handle the specimens, which leads to problems (dropping the meteorite, introducing moisture, oil from the hands, and so on).

- *Riker specimen mounts.* Normally used for the display of botanical and zoological specimens, Riker specimen mounts are very heavy cardboard boxes filled with absorbent cotton that cushions the sample; the top is glass neatly glued to the lid, which is kept in place by four long pins. Riker specimen mounts are an inexpensive and excellent method of presentation and provide good protection for meteorites.
- *Safe.* Some collectors don't take any chances at damaging or losing their meteorites. Specimens are usually placed in Riker specimen mounts, then stored in a safe—fireproof safes recommended.
- *Specimen drawer cabinets.* The serious collector will often have a cabinet with specimen drawers, designed specifically for storing rocks, minerals, and, of course, meteorites. These drawers allow for meteorite storage in a safe and organized manner. A good specimen drawer cabinet, however, can cost $1,000 or more.

A number of collectors use specimen bags, which are much like ziplock baggies, except heavier in weight. Specimen bags

Some sample methods for meteorite storage and display. Top left: Tuxtuac in a Riker specimen mount, an Ordinary Stone Chondrite (LL5). Top right: Red Rock displayed on a plastic stand, an Iron Medium Octahedrite (IIIAB). Left: Brenham in a custom-made wooden case, a Stony-Iron Pallasite (PAL). PHOTO BY DAVID SISSON

Bug box collection of meteorites—affordable, easy to store, and allowing a wide variety of samples. PHOTO BY MIKE MARTINEZ

should be used only on a temporary basis to store newly acquired meteorites. Since specimen bags can be sealed, they can trap moisture with the meteorite, leading to rust. Bug boxes, since they form a fairly tight seal when closed, can also trap moisture, but there is a method to counter that problem.

Desiccant gel, a water absorbing material, can be placed in the bottom of storage boxes to absorb any moisture. Change the desiccant periodically. Desiccant gel is available at several scientific supply houses and occasionally at your local photo supply store.

Once your collection is on display, you should next consider its lighting. Your meteorites should not be displayed in direct sunlight, but artificial lighting can be used to emphasize certain features and characteristics.

A thin section of a pallasite looks simply spectacular when backlit. Do not allow the lights to heat the meteorites.

A good magnifying glass is a useful tool to keep by your collection. You will probably spend a lot of time examining your meteorites.

KEEPING COLLECTION RECORDS

Upon acquiring a new meteorite, *immediately label the specimen.* You can do this in one of three ways: by attaching an adhesive label to the meteorite, by painting an identification number directly on the meteorite, or by labeling the box or stand base with the identification number. Each of these methods has its advantages and disadvantages.

Much information can be included on adhesive labels. However, many meteorite collectors oppose this method because of the label's adhesive glue. Additionally, the labels have a tendency to loose their adhesive over time and fall off the meteorite.

Many collectors, following the example of the late meteorite expert H. H. Nininger, paint a small white strip on the meteorite in an "uninteresting" area. Then, using India ink, an identification number is written on the specimen (e.g., S-99-001, for stony bought in 1999, specimen #1 purchased that year). Other collectors simply paint the number directly onto the meteorite. A few collectors have expressed concern about the possibility that some paints may react chemically with the meteorite; nevertheless, this is the method used by most.

Labeling the box or divided drawer section of a specimen storage cabinet also has its problems. It is always possible to misplace a specimen or inadvertently swap two specimens in the divided drawer, leading to misidentification.

You should get into the habit of keeping a log of your meteorites. The log should contain, at a minimum, the following information:

- *Identification number.* There are several ways to number specimens, including using a sequential series, sequential to the type of meteorite (e.g., PAL001, for the first Pallasite in your collection), or one of your own design.
- *Name of meteorite.* Use the proper fall or find name; confirm the spelling. (Note that there are several meteorites with the same name but a different letter that follows the name; e.g., Pampa A, Pampa B.)
- *Meteorite's classification.*
- *Description*, including gram weight.

METEORITES INVENTORY

Collector: _____

Address: _____

City, State, & Zip: _____

Number	Meterorite; Classification	Weight; Description	Purchased	Dealer; Amount

Recovery of the Goose Lake, California, meteorite, an Iron Medium Octahedrite (IAB), in 1937. Some have suggested the Goose Lake meteorite might actually be a 'thrown-out' of the Meteor Crater, Arizona, impact. PHOTO COURTESY OF THE CHABOT SPACE & SCIENCE CENTER

- *Purchase date.*
- *Purchase amount* (it is also recommended that you keep a copy of the receipt for insurance as well as later resale purposes).
- *Dealer* from whom you purchased the meteorite.

A sample log is found on the previous page.

You might also want to include information about the fall or find's location, date and time that the meteorite was recovered, and interesting information about the meteorite. Many meteorite dealers supply information cards about each meteorite they sell; this is a good starting point.

Many collectors keep a photographic log of their specimens. With today's electronic cameras, a series of images can be processed and easily stored on a CD-ROM.

You may want to check with your insurance company in

regards to coverage in case of fire and theft. Most insurance agents will not be familiar with covering meteorites. However, a special insurance policy rider might be necessary to protect your collection, especially if it contains specimens of high value.

Regardless of whether you collect meteorites as an investment or, as is the case with most of us, because they are special, get to know meteorite dealers you can trust. Protect and catalog your collection, and most of all enjoy your meteorites. They represent a piece of the early solar system, which has remained unmodified and untouched for at least four billion years.

7

Meteorite Craters

Although it is not always possible to own large quantities or even pieces of meteorites, it is possible to visit a number of meteorite craters worldwide. In fact, the earth's best-preserved meteorite crater is in the United States: Meteor Crater, in Arizona.

This chapter provides information regarding craters of interest. Two craters in the United States are highlighted, with additional information on other U.S. craters and impact structures, as well as significant craters worldwide.

You might see the term *astroblem* or *astrobleme* applied to an older, eroded impact structure. The crater form of an impact structure (or astroblemes) has long ago weathered away and is barely recognizable as a meteorite crater.

UNITED STATES

Meteor Crater, Arizona—Latitude 35° 2' N, Longitude 111° 1' W
Located between Winslow and Flagstaff, just south of Interstate 40, Meteor Crater (also referred to as Barringer Crater, Arizona Crater, Canyon Diablo, and Coon Butte or Mountain) has an interesting history, in both its formation and discovery as a meteorite crater. Approximately fifty thousand years ago, an iron meteorite 150 feet across, traveling at thirty to forty thousand miles per hour, slammed into earth. At impact, the main meteorite mass

Crater cross section showing the overall distribution of material. In the formation of a crater the size of Meteor Crater, at the point of impact almost all the rock is destroyed. Other rock and material is thrown outside the crater and into the atmosphere. Rock at the rim is overturned and some material from the impact fills the crater. ILLUSTRATION BY DAVID FRANTZ

mostly vaporized, showering the area with thousands of pounds of meteorite fragments. The incredible pressure at impact, estimated to have been more than twenty thousand times greater than normal atmospheric pressure, formed two types of silica as well as microscopic diamonds. The impact crater is now 570 feet deep and 4,100 feet across. It has been estimated if we were to use the crater like a football or soccer stadium, it would seat one million people. Meteor Crater is the earth's best example of a meteorite crater; it is in pristine condition with little erosion.

The discovery that Meteor Crater was formed by a meteorite impact had to wait until the twentieth century. The crater had been known to Native Americans living in the area, but the first written report was made in 1871 by a scout for General Custer; the crater was initially called Franklin's Hole after the scout. Local settlers later named it Coon Butte, thinking it was an extinct volcano.

In the 1890s the first scientist visited Meteor Crater. Dr. A. E.

An aerial view of Arizona's Meteor Crater. PHOTO BY DR. DAN DURDA

Foote, a chemist and mineralogist from Philadelphia, returned with a large quantity of Meteor Crater meteorites. Foote's investigation uncovered the tiny diamonds within some of the meteorites, formed under tremendous pressure. His paper on Meteor Crater caught the attention of U.S. Geological Survey chief geologist Grove Karl Gilbert. However, Gilbert thought the crater had been formed by volcanic means, even with the discovery of meteorite fragments around it.

A mining entrepreneur by the name of Daniel Moreau Barringer spent much of his life and personal wealth exploring the crater, looking for the main meteorite mass, and convincing others that Meteor Crater was formed by the impact of a meteorite.

Barringer attempted to drill several holes to reach, what he believed, was the buried main meteorite mass. Several of the holes were over one thousand feet deep, but jammed and broken drills, as well as a flooded shaft, ended the quest to mine what Barringer envisioned as the meteorite's iron and nickel content. Even though he never recovered "the main meteorite mass," he did find diamonds, platinum, and iridium in addition to meteorite fragments. The largest meteorite recovered was 639 kilograms. Barringer authored several papers about Meteor Crater as a result of his investigation and research.

The argument raged on over the Crater's formation. It wasn't until 1930 that most geologists and astronomers accepted the fact that the Crater was formed by a meteorite impact.

The acceptance that Meteor Crater was caused by meteorite impact lead to the immediate identification of several other terrestrial craters, including the Henbury Craters in Australia and the Odessa Craters in Texas.

Meteor Crater's history includes many extensive studies, including those by H. H. Nininger (who established a small museum off Route 66, which is now Interstate 40, near the crater) and Eugene Shoemaker. Meteor Crater was used by NASA to train the Apollo astronauts. These training sessions, initially conducted by Shoemaker, assisted the astronauts in preparation to explore the moon.

A small museum called the Museum of Astrogeology, operated by Meteor Crater Enterprises, has been established at the crater's rim. Meteor Crater was designated a Natural Landmark

View of Meteor Crater from the north rim. PHOTO BY DR. MIKE REYNOLDS

The remains of H. H. Nininger's American Meteorite Museum in the foreground, with Meteor Crater seen to the right of the Museum on the horizon. PHOTO BY DR. MIKE REYNOLDS

in 1968 by the U. S. Secretary of Interior. Note that collecting of any type is not allowed at or around the crater.

Odessa, Texas—Latitude 31° 45' N, Longitude 102° 29' W

Just southwest of Odessa lies a series of meteorite craters. The main crater is about 550 feet across and exhibits a slightly raised rim. The area has been heavily excavated, but the craters are still recognizable as such, even though they have suffered years of neglect and abuse. Tons of coarse octahedrites have been removed from the Odessa craters, but you can still find meteorites at the Odessa site. Beware that the craters are on grazing property. Cattle rustling has been a problem in the area so it's a good idea to clearly identify yourself as a meteorite enthusiast.

Daniel Moreau Barringer's son D. Moreau Barringer, Jr., proved in 1926 that the Odessa craters were caused by meteorite impacts.

View of the Odessa, Texas, crater area. PHOTO BY MIKE MARTINEZ

Beaverhead, Montana—Latitude 45° 0' N, Longitude 113° 0' W

This partially exposed and eroded impact structure was first recognized as such in 1990 because of the discovery of shatter cones at the site. *Shatter cones* are formed when impact shock waves produce weakened areas on surrounding rock, which erodes into a ridged, cone-shaped structure. The impact structure is located in southwest Montana at its border with Idaho.

Crooked Creek, Missouri—Latitude 37° 50' N, Longitude 91° 23' W

A partially exposed impact structure identified by the presence of shatter cones. The structure is southeast of Rolla and south of Steelville; Crooked Creek runs through the structure's eastern side.

Decaturville, Missouri—Latitude 37° 54' N, Longitude 92° 43' W

A partially exposed impact structure identified by the presence of shatter cones common in the crater's center. The structure is located just south of Camdenton, and Decaturville lies on the northeast rim.

Flynn Creek, Tennessee—Latitude 36° 17' N, Longitude 85° 40' W

Shatter cones and other evidence of impact have been found in this partially exposed and eroded impact structure. The structure is located about twenty miles north of Cookeville and is centered on the town of Clenny in northern Tennessee.

Glover Bluff, Wisconsin—Latitude 43° 58' N, Longitude 89° 32' W

This impact structure was first identified as such in 1983 with the discovery of shatter cones. Quarries have been established at the site for the mining of dolomite. There are three hills at the site, located north of the Wisconsin Dells and the town of Lawrence.

Haviland, Kansas—Latitude 37° 35' N, Longitude 99° 10' W

This fresh, but unrecognizable crater has produced thousands of pounds of Brenham meteorites, a pallasite. Most of these meteorites are highly weathered and oxidized. H. H. Nininger first recognized the crater in 1925. Much wear on the crater has been due to its excavation for use as a livestock watering hole. The site is located east of Greensburg and a couple of miles south of US Highway 54.

Marquez Dome, Texas—Latitude 31° 17' N, Longitude 96° 18' W

This fairly young, partially exposed, large impact structure (about nine miles across) features shatter cones at its center, as well as other impact qualities. The structure is located near the city of Marquez.

Middlesboro, Kentucky—Latitude 36° 37' N, Longitude 83° 44' W

This old and highly eroded structure a little less than four miles across, is easy to identify as an impact structure because it is flat, relatively speaking, compared with the surrounding topography. The structure is easy to locate because the city of Middlesboro is located on the site.

Serpent Mound, Ohio—Latitude 39° 2' N, Longitude 83° 24' W

This impact structure, a little more than three and a half miles across, is characterized by shatter cones. Serpent Mound lies

Shattercone from the Sierra Madera, Texas, impact feature. PHOTO BY DR. MIKE REYNOLDS

below Native American earthworks and includes Serpent Mound State Memorial. The structure is located southwest of Sinking Springs in the southern part of the state.

Sierra Madera, Texas—Latitude 30° 36' N, Longitude 102° 55' W

This unusual formation does not appear to the casual observer to be an impact structure. However shatter cones and many other impact features, such as shocked quartz, are present. The structure is located about eighteen miles south of Fort Stockton.

Upheaval Dome, Utah—Latitude 38° 26' N, Longitude 109° 54' W

Confirmed as an impact structure in 1984, Upheaval Dome is highly eroded and partially exposed. A central uplift can be seen. Located in Canyonlands National Park, Upheaval Dome is easily accessed via a Park Service road.

Versailles, Kentucky—Latitude 38° 2' N, Longitude 84° 42' W

Versailles is believed to be an impact structure, but on less evidence than other established structures. It is highly eroded and only partially exposed. The impact structure is three miles north-east of Versailles; Big Sink Road passes almost directly through it.

CANADA

Brent, Ontario—Latitude 46° 5' N, Longitude 78° 29' W

Located near the town of Deux Rivieres, this is a partially exposed impact structure about 2 ⅓ miles across.

Carswell, Saskatchewan—Latitude 58° 27' N, Longitude 109° 30' W

A large structure that is partially exposed, Carswell is located in a remote part of Canada.

Charlevoix, Quebec—Latitude 47° 32' N, Longitude 70° 18' W

The Charlevoix impact structure seems to come out of the St. Lawrence River. This large structure, of which only about half is visible, is about thirty miles across at its widest point. Charlevoix is near Quebec in the Charlevoix Region along Provincial Highway 138, which follows the St. Lawrence River.

Clearwater Lakes, Quebec—Latitude 56° 13' N, Longitude 74° 30' W

These two lakes of unequal size were formed by a double impact. They are clearly impact structures, evidenced by the recovery of shatter cones and other impact characteristics. The Clearwater Lakes are about sixty miles inland from Hudson Bay's Eastern Shore in northern Quebec and are accessible by float plane.

Haughton, Northwest Territories—Latitude 75° 22' N, Longitude 89° 40' W

For the adventurer, the Haughton impact structure, also known as the Haughton Dome, represents a nice opportunity. This fairly large crater features shatter cones and other impact structure characteristics. Haughton is on Devon Island in the Canadian Arctic.

The Clearwater Lakes twin impact structures as seen from earth orbit. PHOTO TAKEN DURING SPACE SHUTTLE MISSION STS 61A, NASA

Holleford, Ontario—Latitude 44° 28' N, Longitude 76° 38' W

This 1.4-mile diameter, old, partially exposed crater is now a depression with the crater's rim barely visible. The center of the crater is wet almost year-round. Holleford is easily accessible; it lies about seventeen miles northwest of Kingston.

New Quebec, Quebec—Latitude 61° 17' N, Longitude 73° 40' W

From the air, the New Quebec crater looks very similar to Meteor Crater in Arizona, except it is larger, with a diameter of a little over two miles (versus 4,100 feet for Meteor Crater). New Quebec is filled with water but has retained a clearly identifiable crater rim. The crater is in a remote area of Ungula Peninsula and can only be reached via float plane.

OTHER NOTABLE CRATERS WORLDWIDE

The following is a listing of the major earth impact craters or craters of unique interest. They are listed alphabetically by country.

Amguid, Algeria—Latitude 26° 5' N, Longitude 4° 23' E

A relatively young crater, Amguid Crater is about 1,475 feet in diameter. It is not located near a major highway and is about 140 miles southeast of Oasis of In Salah.

Campo Del Cielo Craters, Argentina—Latitude 27° 38' S, Longitude 61° 42' W

A total of twelve craters make up Campo Del Cielo. The largest is only about 320 feet across, the smallest about 65 feet. Numerous meteorites, by the same name as the craters, have been recovered.

Boxhole Crater, Northern Territory, Australia—Latitude 22° 37' S, Longitude 135° 12' E

This slightly eroded crater is a little larger than six hundred feet in diameter. Medium octahedrite meteorites have been collected from the surrounding area. The closest major city is Alice Springs, about 165 miles away.

Henbury Craters, Northern Territory, Australia—Latitude 24° 35' S, Longitude 133° 9' E

A total of thirteen craters make up the Henbury impact area. The largest crater is elliptical, about 460 feet by 590 feet. The smallest crater is only about 20 feet across. At least one crater is no longer visible and its existence is debated. The Henbury Craters are protected as a Conservation Reserve. The craters are about ninety miles from Alice Springs.

Wolfe Creek, Western Australia, Australia—Latitude 19° 18' S, Longitude 127° 47' E

A well-preserved crater, Wolfe Creek is the most famous impact crater in Australia. It is similar in structure to Meteor Crater in Arizona, although a little smaller at about 2,890 feet across. Wolfe Creek is in a national park, and was created with the impact of an octahedrite. The crater is accessible by an unpaved road, about a seventy-five mile drive south-southwest of Halls Creek.

Kaalijarvi Craters, Estonia—Latitude 58° 24' N, Longitude 22° 40' E

A total of seven craters in a geological preserve have been identified as a part of the Kaalijarvi Craters. The main crater, filled with water, is about 360 feet across. Medium octahedrites have been found at the site. Kaalijarvi Craters are on the island of Saarimaa.

Ries, Germany—Latitude 48° 53' N, Longitude 10° 37' E

The Ries impact crater is an example of a large, old impact structure. Ries Crater is nearly fifteen miles across, flat in structure, with a central uplift. Characteristics of an impact, including shatter cones, have been found in the area. The old walled city of Nordlingen lies in the structure's basin; an excellent museum is located there.

Aouelloul, Mauritania—Latitude 20° 15' N, Longitude 12° 41' W

A fairly young crater with a diameter of a little over 1,300 feet, Aouelloul is interesting because of the impact glass found in and around the crater. The crater is thirty miles southeast of Atar.

Morasko Craters, Poland—Latitude 52° 29' N, Longitude 16° 54' E

A total of seven craters have been identified as part of the Morasko Craters, the largest about 325 feet across and the smallest a little less than 50 feet in diameter. There may have been additional craters, but farming would have altered these. Coarse octahedrite meteorites have been recovered from the surrounding land. A meteorite sanctuary, the Morasko Craters are just west of the village of Morasko.

Sikhote-Alin, Russia—Latitude 46° 7' N, Longitude 134° 40' E

Numerous craters and holes dot the site of this impact, which occurred on February 12, 1947. The largest crater of at least 120 individual craters formed is just less than ninety feet in diameter. The octahedrite meteorites resemble shrapnel. Thousands have been recovered, including meteorites stuck in trees. Sikhote-Alin is in the western foothills of the Sikhote-Alin Mountains, just east of the town of Guberovo and several hundred miles north of Vladivostock.

8

Tektites

Tektites are fused glassy material classified as "dry"—that is, they contain very little water. They are silica-rich (approximately sixty-five to eighty percent silica, or SiO_2), and somewhat similar in composition to volcanic glasses such as obsidian. However, the external appearance of tektites is considerably different from that of obsidian. Tektites may be black, gray, brown, and even a translucent green. They appear to have been melted, and their shapes indicate rapid flight through the earth's atmosphere at some time during their formation.

TEKTITE HISTORY AND ORIGINS

Tektites have been used as ornaments, amulets, and tools, some of which date back six thousand to eight thousand years. Tektites were worn as good luck charms after the Iron Age, around 500 B.C. Liu Sun in China wrote the first reference to tektites in A.D. 950, calling them *Lei-gong-mo*, meaning "inkstone of the Thundergod."

Joseph Mayer was referring to tektites in 1788, when he described them as a type of volcanic glass. Another theory from the eighteenth century described tektites as shards of hand-made glass. Yet another stated that the Aboriginals made the Australian tektites; however, the Aboriginals had their own theories about the origins and mystical powers of the stones.

In the nineteenth century, other theories about tektite origins were put forth. In 1844, Charles Darwin, the English naturalist, classified tektites as plain obsidian. In 1893, Victor Streich, a geologist, stated that tektites were a form of meteorite, and in 1897, Dutch geologist R. D. M. Verbeek theorized that the origin of tektites was lunar volcanoes.

F. E. Seuss, an Austrian geologist, was the first to use the term "tektite" in 1900. It comes from the Greek word *tektos*, which means melted or molten. Seuss stated that tektites were extraterrestrial in origin, a type of glass meteorite. Seuss also believed that tektites got their unusual shapes due to high-velocity airflow. Because of Seuss' work, many universities and museums grouped tektites with meteorites.

H. H. Nininger, the great meteorite collector, proposed in 1940 that tektites were the result of meteorite impacts on the moon. According to Nininger, lunar "splash" or ejecta material from the impacts were making it to the earth.

Since no one has ever witnessed a bright meteor or fireball from which tektites have been recovered, tektite origin as glassy meteorites has been generally dismissed. There are a few scientists who continue to support the theory that tektites are due to earth volcanism, which produces obsidian glass. The problem with this theory is that the earth's volcanoes are incapable of ejecting materials high enough into the atmosphere to allow for tektites' unusual shaping, which many believe occurs during high-velocity entry through the atmosphere. Additionally, the chemical composition of tektites is significantly different from that of volcanic glasses such as obsidian.

The lunar volcanism theory suggests that lunar volcanoes ejected the glassy material, which eventually entered the earth's atmosphere and underwent shaping. There are also reliable challenges to this theory. First of all there is no evidence of recent lunar volcanism. Second, and most problematic for the volcanism theory, is the fact that lunar materials returned to the earth from American Apollo and Soviet Luna expeditions are significantly different from tektites in chemical composition.

Two other theories are currently under study. Both of these

focus on the elements of *molten silicate* and *high-velocity shaping*. And both theories are based on extraterrestrial impacts.

According to the theories, tektites are formed when a meteorite or comet strikes the earth or the moon with such force that the surrounding rock is vaporized and ejected into space. In space the ejecta droplets begin to cool and harden, and then begin their fall back toward the earth, getting their high-velocity shaping as they enter the earth's atmosphere. These theories differ only in whether tektites are considered ejecta from meteorite/comet impacts on the earth or on the moon.

Dr. John O'Keefe at NASA's Goddard Space Flight Center supported the theory that tektites originated from lunar impacts. It is well documented that the moon has been bombarded time and time again. And lunar meteorites have been found on the earth. According to some researchers, the *shape* of one specific tektite group—those from Australia—indicate they may be lunar ejecta. Tektites probably take time to form, thus better supporting the lunar-impact over the earth-impact theory. The major weakness of the lunar-impact-ejecta theory comes from the study of lunar materials returned to the earth from Apollo and Luna missions. And if tektites are of lunar origin due to meteorite or comet impact, why are the locations of tektites on the earth restricted to only four areas?

The theory that tektites are meteorite or comet earth-impact ejecta is by far the most plausible and widely accepted theory, and there is a good deal of evidence to support it. Traces of meteoritic iron-nickel have been found in some Asian tektites. Analyses of rare elements found in tektites provide additional evidence of the terrestrial meteorite impact theory. Furthermore, it has been shown that impact ejecta move at a higher velocity than the incoming meteorite or comet.

But the meteorite or comet earth-impact ejecta theory also has its problems. Foremost is the velocity required for impact ejecta to be shot up through the earth's atmosphere and into space. Would an impact deliver sufficient energy for this velocity to be reached?

The ages of tektites are a problem for all of these theories. It appears that tektites were formed only over the last thirty-five million years (the age of the oldest tektites found in Texas). Why don't tektites exist from earlier time periods?

COMPOSITION

Tektites are akin to earth glass, in particular to obsidian. The primary difference between tektites and earth glass is the very low water content of tektites, which also exhibit a low alkali content, and contain pure silica glass (called *lechatelierite*), coesite (a dense silica polymorph), a different form of iron than earth glasses, and baddeleyite, a zircon oxide (ZrO_2) mineral produced at high temperatures. (Consult books and articles about tektites for detailed chemical analysis.)

TEKTITE FORMS

Beginning as a molten blob of glass in the earth's atmosphere, a tektite begins to solidify and take on a primary form. Tektites are found in four basic forms, which are dependent on the velocity at which the original molten material enters the atmosphere.

Splash Form. The most common form of tektite is the splash form, which is homogeneous in composition. Splash form tektites are shaped by rotation that occurs during entry of the molten material into the earth's atmosphere. They take on four basic shapes, which depend on the speed at which the material is rotating:

- *Spheres* appear to have been formed when the ejecta material did not rotate.
- *Ellipsoids* and *spheroids* are formed by slow to moderate rotation; these range from elliptical to bar-shaped tektites.
- *Dumbbells* are formed by rapid rotation; the bar-shaped tektite continues to widen at opposite ends and shrink in the middle.
- *Apioids* or *teardrops* are formed in a way similar to the dumbbell shaped tektites, except the speed of rotation was so fast that the dumbbell separated, creating two teardrop-shaped tektites that underwent little or no additional rotation.

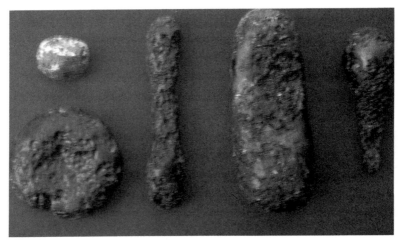

Splash form Indochinite tektite shapes with weights from 24 to 214 grams. PHOTO BY MIKE MARTINEZ

The most common shapes are the spheres; well over half to two-thirds of tektites found are spheres. About a quarter of tektites are ellipsoids and spheroids. Dumbbells make up less than ten percent and apioids less than half that number. A number of other shapes derived from these four basic splash-form shapes are also found, including bars, boat shapes, and disks.

Layered. These basically shapeless tektites, made up of thin, heterogeneous layers, are primarily found in Southeast Asia and are also known as Muong Nong–type tektites.

Ablated. These aerodynamically shaped tektites probably started as splash-form tektites but with little or no rotation. Traveling farther and faster than the splash-form tektites, ablated tektites may have gotten their unique shape by not tumbling or spinning while reentering. These tektites experienced a significant atmospheric drag as they entered the atmosphere. The side facing the atmosphere formed differently than the side facing away from the earth. Ablated tektites have only been found in Australia and come in two shapes, called cores and buttons.

Microtektites. This final form of tektite is small (less than

Assorted tektites. Left: Maom Ming City, 9 grams. Middle: Phillippin-
ite, 17 grams. Right: Australian button, 3 grams. PHOTO BY MIKE MARTINEZ

one millimeter in diameter) and usually found on the ocean floor
(with the exception to date of the island of Barbados). In all other
aspects—age, composition, and shape—they are akin to other
tektites.

ADDITIONAL SHAPING

After the primary formation occurred, additional shaping or
sculpturing also happened. Due to its newly formed shape, the
tektite stayed in basically the same orientation during its plunge
through the atmosphere until it slowed down. The area of the tektite directly exposed to the earth's atmosphere again heated up,
forming pits. The ablated tektites acted similar to the heat shield
on one of our earlier spacecraft. A layer of the tektite ablated—
that is, melted and flowed away in a stripping manner, forming
flow lines. Additional sculpting took place depending on the size
of the falling tektite and once on the ground, weathering and erosion began, further shaping the tektite. Tektites have survived the
process of weathering profoundly better than meteorites.

WHERE TEKTITES ARE FOUND

Unlike meteorites, which are found worldwide, tektites are found in four major geographical areas or *strewn fields*. These are in North America, Czechoslovakia, the Ivory Coast of Africa, and Southeast Asia and Australia.

At least three additional strewn fields are occasionally referenced; these usually refer to other classes of terrestrial glasses.

NORTH AMERICAN TEKTITES

About thirty-five million years old, North American tektites are found in two primary locales in Texas and Georgia. A single tektite recovered in 1959 from Martha's Vineyard, in Massachusetts, and a second tektite reportedly from Cuba are considered by some to be anomalies.

Bediasites, named for the city of Bedias, were first discovered in Texas in 1939. They can be found in a narrow strip of land that parallels the Gulf Coast of Texas and is located about ninety miles inland in the counties of De Witt, Gonzales, LaVaca, Lafayette, Lee, Burleson, Brazos, Grimes (location of Bedias), and Walter

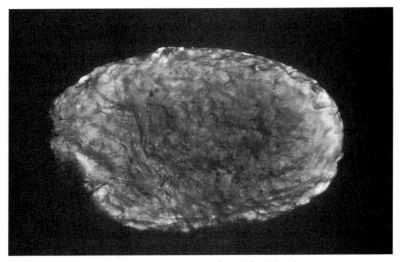

A 28.1-gram Georgia tektite, recovered by Mike Martinez. PHOTO BY DR. MIKE REYNOLDS

(from the southwest to the northeast). Bediasites are spherical or half-spherical in shape, usually quite weathered and relatively smooth, and dark brown or brownish green.

First found in 1938, olive green georgiaites are primarily from southeast Georgia, centered in Dodge and Bleckley counties, though a couple of georgiaites have recently been found in areas considerably outside the recognized strewn field. There are far fewer georgiaites than bediasites.

CZECHOSLOVAKIAN TEKTITES

Tektites from the Czech Republic and Austria are known as *moldavites*. They are found in Moravia and Bohemia and were first discovered in the late eighteenth century. Characteristically a pale green color, moldavites are flattened disks with irregular indented edges. Moldavites are about 14.7 million years old. The largest moldavite recovered to date weighs close to five hundred grams, much larger than the North American tektites, which weigh up to ninety grams, or Ivory Coast tektites.

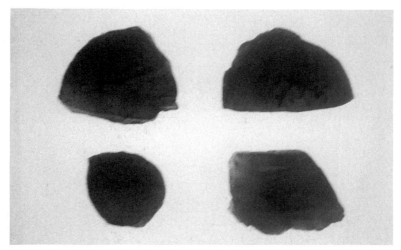

Moldavite tektites. When backlit as with these 7- to 15-gram Moldavites, they show a beautiful greenish hue, among other characteristics. PHOTO BY MIKE MARTINEZ

IVORY COAST TEKTITES

The tektites from the Ivory Coast of Africa, sometimes referred to as ivorites, are about 1.3 million years old. The largest Ivory Coast tektite weighs only about eighty grams. Only a few hundred have been recovered, most because of gold mining.

SOUTHEAST ASIAN AND AUSTRALIAN TEKTITES

Often referred to as the Australasian Strewn Field, this area includes Australia and New Zealand, Cambodia, southern China, Laos, Thailand, Vietnam, Java, Malaysia, and the Philippines. By far the largest in size, and claiming the greatest variety and number of tektites, the Australasian Strewn Field also appears to be the youngest at about seven hundred fifty thousand years old.

The variety and availability of Australasian Strewn Field tektites make these tektites a great resource for beginners to start their collection. Many of the indochinites are available at affordable prices, but the finer, museum-quality specimens will cost more. The beautiful Australian buttons are very expensive. Be prepared to pay hundreds of dollars for a complete button sample and thousands for a museum-quality piece.

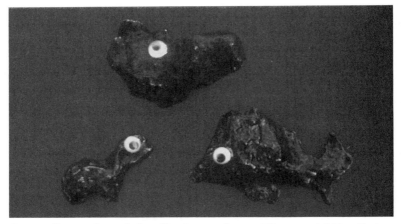

Some collectors base their collection on shapes. These are tektites that are shaped like animals—with an added feature. Top: dog's head. Bottom left: turtle. Bottom right: fish. PHOTO BY MIKE MARTINEZ

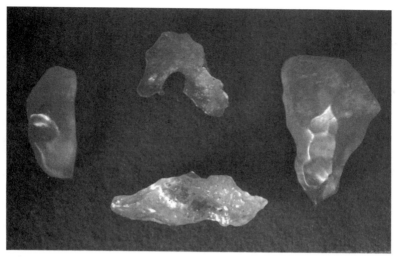

Libyan Desert Glass, a naturally formed glass not considered to be a tektite. PHOTO BY DR. MIKE REYNOLDS

STORAGE AND DISPLAY OF TEKTITES

Tektites can be stored and displayed in the same manner as meteorites. Some tektites may have fragile edges (prone to damage), so you should use caution in handling them. Riker specimen mounts are an excellent way to display and store your tektites.

Back-lighting green tektites, such as the moldavites and Georgia tektites, will produce a beautiful display. Again, as with meteorites, use caution in regards to the heat from the light.

OTHER NATURAL GLASSES

There are a number of other glasses often grouped with tektites. Many of these are created by other natural phenomena. Lightning strikes can cause soil fusion, creating a braided tube of glass up to three feet long, called a fulgurite. Volcanic glass, such as obsidian, is one of the main products of acid volcanism (magma with a high silica content). There are other natural glasses, such as Libyan Desert Glass and Darwin Glass from Tasmania, that are not considered tektites.

Appendix A:
Bibliography

A number of references are listed below. Broken into several categories—including Children's Books and General Skywatching and Astronomy—you will find a number of excellent choices for further reading. Some of these books might not be available from your local new bookseller so try a used bookstore or an on-line search. Good reading!

METEORS

Bone, Neil. *Meteors: A Sky & Telescope Guide*. Cambridge, Mass.: Sky Publishing, 1993. *A guide to observing, reporting, and photographing of meteors.*

Edberg, Stephen J., and David H. Levy. *Observing Comets, Asteroids, Meteors and the Zodiacal Light*. Cambridge: Cambridge University Press, 1994. *An excellent guide by two highly respected astronomers.*

Kronk, Gary W. *Meteor Showers: A Descriptive Catalog*. Hillside, N.J.: Enslow Press, 1988. *An excellent documentation of meteor showers and histories.*

Littmann, Mark. *The Heavens on Fire*. Cambridge: Cambridge University Press, 1998. *The author explores the history and science of meteors and the great Leonid storms in particular.*

Lunsford, Robert D. *The A.L.P.O. Guide to Watching Meteors*. Washington, D.C.: The Astronomical League, 1995. *An introduction to meteor watching; includes forms for visual observations.*

Olson, Roberta J. M., and Jay M. Pasachoff. *Fire in the Sky*. Cambridge: Cambridge University Press, 1998. *A beautifully illustrated overview of British science and art in the area of comets and meteors.*

METEORITES

Bagnall, Philip M. *The Meteorite & Tektite Collector's Handbook*. Richmond: Willmann-Bell, 1991. *A guide specifically geared to collecting, preserving, and displaying meteorites with numerous photographs to illustrate the text.*

Burke, John G. *Cosmic Debris: Meteorites in History.* Berkeley and Los Angeles: The University of California Press, 1986. *A historical overview of meteorite studies, including some contemporary work.*

Carion, Alain. *Meteorites.* Paris: Self-published. *Translated into English, this book is a nice overview of meteorites.*

Desonie, Dana. *Cosmic Collisions.* New York: Henry Holt and Company, 1996. *From* Scientific American's Focus *series, overview impacts and historical solar system collisions, including the potential of a future earth impact.*

Dodd, Robert T. *Meteorites: A Petrologic-Chemical Synthesis.* Cambridge: Cambridge University Press, 1981. *An advanced study of meteorites.*

———. *Thunderstones and Shooting Stars: The Meaning of Meteorites.* Cambridge, Mass.: Harvard University Press, 1986. *A general reference source; one of the best for those looking to increase their knowledge about meteorites.*

Heide, Fritz, and Frank Wlozka. *Meteorites: Messengers from Space.* Berlin: Springer-Verlag, 1994. *An updated translation of Heide's 1957* Kleine Meteoritenkunde, *an overview of meteorites.*

Grady, Monica M. *Catalogue of Meteorites.* 5th ed. Cambridge: Cambridge University Press, 2000. *The current edition of this outstanding reference catalog includes a CD-ROM listing for PC's.*

Graham, A. L., A. W. R. Bevan, and R. Hutchison. *Catalogue of Meteorites.* 4th ed. Tucson: The University of Arizona Press, 1985. *A reference catalog of meteorites, including the location, find or fall date, amount recovered as known, classification, and a short description of each meteorite.*

Hutchison, Robert. *The Search for Our Beginning.* Oxford: Oxford University Press, 1983. *Overviews meteorites historically and scientifically, with an emphasis on solar system origin and development.*

Hutchison, Robert, and Andrew Graham. *Meteorites.* New York: Sterling Publishing Company, 1993. *A general overview and introduction to meteorites.*

Jensen, Michael R., William B. Jensen, and Anne M. Black. *Meteorites from A to Z.* Aurora, Colo.: Michael R. Jensen Publisher, 2001. *A complete listing of all 3,317 named meteorites as of December 31, 2000, with the exception of the unnumbered Antarctica meteorites.*

Matthews, Kerridge, and Mildred Shapley Matthews, eds. *Meteorites and the Early Solar System.* Tucson: The University of Arizona Press, 1988. *A large reference written by 69 collaborators.*

McSween, Jr., Harry Y. *Meteorites and Their Parent Planets.* 2nd ed. Cambridge: Cambridge University Press, 1999. *A thorough overview of the science of meteorites.*

Morgan, Matthew L. *The Handbook of Colorado Meteorites.* Denver: Colorado Geological Survey, 2000. *An example of a small, but very useful identification guide to meteorites.*

Nininger, Harvey H. *Ask a Question about Meteorites.* Denver: American

Meteorite Laboratory, 1989. *This wonderful little book, once available readily in museums and science centers, provides a good overview of meteorites in a question-and-answer format.*

————. *A Comet Strikes the Earth.* Denver: American Meteorite Laboratory, 1969. *This classic Nininger book is actually about meteorites and, in a circular cut-out, included a Meteor Crater, Arizona, meteorite chip.*

————. *Find a Falling Star.* New York: Paul S. Eriksson, 1972. *Nininger's classic should be on every serious meteorite collector's shelf; the book might be difficult to find, but is worth the effort.*

Norton, O. Richard. *Rocks from Space.* 2nd ed. Missoula, Mont.: Mountain Press Publishing, 1998. *An extensive guide to meteorites, one which must be on every meteorite collector's bookshelf.*

Pejovic, Brian. *Man and Meteorites.* Avening Tetbury Gloucestershire, United Kingdom: Thames Head, 1982. *An introduction to meteorites.*

Regelman, Kenneth. *ARN's History of Meteorites.* Maplewood, Minn.: Astronomical Research Network, 1997. *This reference lists over 3,000 meteorites, including the location, find or fall date, classification, and a short description of each meteorite.*

Smith, Dean. *The Meteor Crater Story.* Flagstaff, Ariz.: Meteor Crater Enterprises, 1996. *An overview of the formation, exploration, and preservation history of Meteor Crater.*

Steel, Duncan. *Target Earth.* Pleasantville, N.Y.: The Reader's Digest Association, 2000. *An overview of major planetary impacts, evidence of such impacts, and earth-based searches.*

Wasson, John T. *Meteorites.* New York: W. H. Freeman, 1985. *A technical book dealing with the origins of the various meteorite types.*

Willey, Richard R. *The Tucson Meteorites: Their History from Frontier Arizona to the Smithsonian.* Tucson: The University of Arizona Press, 1997. *An overview of the fascinating history of two meteorites, one of which is the classic Tucson Ring now on exhibition at the Smithsonian Institution's National Museum of Natural History.*

METEORS AND METEORITES

Knight, David C. *Meteors and Meteorites.* New York: Franklin Watts, 1969. *A good introductory book on the subject, covering several historical aspects of meteors and meteorites.*

Povenmire, Harold R. *Fireballs Meteors & Meteorites.* Indian Harbour Beach, Fla.: JSB Enterprises, 1980. *The author explores many of his personal observing and collecting experiences.*

Sumners, Carolyn, and Carlton Allen. *Cosmic Pinball.* New York: McGraw-Hill, 2000. *A current reference which overviews the latest scientific research on comets, meteors, and asteroids.*

CRATERS AND IMPACTS

Alvarez, Walter. *T. rex and the Crater of Doom.* Princeton, N.J.: Princeton University Press, 1997. *Written by one of the two scientists who proposed the Yucatan Peninsula Chicxulub crater formation by a large asteroid that* lead to the end of the dinosaur era, this book takes the reader through the theory's compelling evidence as developed by Alvarez and his father.

Davies, John K. *Cosmic Impact*. New York: St. Martin's Press, 1986. *This resource details catastrophic earth impacts.*

Hodge, Paul. *Meteorite Craters and Impact Structures of the Earth*. Cambridge: Cambridge University Press, 1994. *A guide to craters worldwide, including location and specifics about each known crater.*

Hoyt, William G. *Coon Mountain Controversies: Meteor Crater and the Development of Impact Theory*. Tucson: The University of Arizona Press, 1987. *A great overview of Meteor Crater history.*

LeMaire, T. R. *Stones from the Stars*. Englewood Cliffs, N.J.: Prentice-Hall, 1980. *The author reviews specific meteorite falls and craters.*

Mark, Kathleen. *Meteorite Craters*. Tucson: The University of Arizona Press, 1987. *An overview of crater formation, a history of crater formation exploration and theories, and craters around the world.*

Nininger, H. H. *Arizona's Meteorite Crater*. Denver: American Meteorite Laboratory, 1956. *Another book—if you can find a copy—by the famous meteorite collector providing a solid overview of Meteor Crater in Arizona.*

Peck, Bill. *Guide to North American Meteorites*. Steamboat Springs, Colo.: self-published, 2000. *An excellent map showing locations of North American meteorites falls and finds.*

Peebles, Curtis. *Asteroids*. Washington, D.C.: Smithsonian Institution Press, 2000. *An overview of asteroids, their formation and types, asteroid impacts in the solar system, including the Shoemaker-Levy 9 impact into Jupiter, and possible earth impacts.*

Tobin, James P. *Meteor Crater: The First Hundred Years of Exploration*. Self-published, 1998. *An overview of Arizona's Meteor Crater, including the history of exploring the crater.*

Verschuur, Gerrit L. *Impact! The Threat of Comets and Asteroids*. New York: Oxford University Press, 1996. *An overview of the effects of impacts on the earth, the search for such potentially devastating objects, and preventive measures for such impacts.*

TEKTITES

Barnes, Virgil E., and Mildred A. Barnes, eds. *Tektites*. Stroudsburg, Pa.: Dowden, Hutchinson & Ross, 1973. *A highly technical selection of research papers on the subject of tektites.*

Heinen, Guy. *Tektites*. Luxembourg: Self-published, 1998. *A rich resource regarding specific tektites, with various collectors' notes, comments, and references.*

McNamara, Ken, and Alex Bevan. *Tektites*. Rev. ed. Perth, Australia: Western Australian Museum, 1991. *An overview of tektites and their origins, focusing on australites.*

O'Keefe, John A., ed. *Tektites*. Chicago: The University of Chicago Press, 1963. *A highly technical, older book edited by one of the experts on tektites.*

Povenmire, Harold R. *Tektites: A Cosmic Paradox*. Indian Harbour Beach, Fla.: Self-published, 1998. *This book is an excellent overview of Georgia tektites in particular, including many collecting stories as well as the tektite origin debate.*

CHILDREN'S BOOKS ABOUT METEORITES

Kraske, Robert. *Asteroids: Invaders from Space.* New York: Aladdin Paperbacks, 1995. *Overviews asteroids, as well as earth impacts.*

Meteor Crater. La Jolla, Calif.: Positive Publishing, 1999. *An activity and coloring book geared for young children.*

Nicolson, Cynthia Pratt. *Comets, Asteroids and Meteorites.* Toronto: Kids Can Press, 1999. *A nicely illustrated children's book with activities.*

Rosen, Sidney. *Can You Catch A Falling Star?* Minneapolis: Carolrhoda Books, 1995. *A richly illustrated book that overviews meteors and meteorites.*

Sipiera, Paul P. *Meteorites.* Chicago: Children's Press, 1994. *A great book for young children, richly illustrated with color photographs.*

Zim, Herbert S. *Shooting Stars.* New York: William Morrow and Company, 1958. *This classic children's book is one that many adult meteor enthusiasts remember checking out from their school library and reading. Unfortunately, many schools have taken it off the shelf because of its age. If you can get a copy, do so!*

GENERAL SKYWATCHING AND ASTRONOMY

Bishop, Roy L., ed. *Observer's Handbook.* Toronto: Royal Astronomical Society of Canada, 1909–. *An annual reference featuring a daily calendar of astronomical events; includes meteor showers and meteorite craters.*

Burnam, Robert, et al. *Advanced Skywatching.* New York: Time-Life Books, 1997. *A solid guide of high-level observing projects and techniques.*

Covington, Michael A. *Astrophotography for the Amateur.* 2nd ed. Cambridge: Cambridge University Press, 1999. *A guide to all areas of astronomical photography.*

Discovery Channel. *Night Sky.* New York: Discovery Books, 1999. *An introductory guide to the night skies, including constellation overviews.*

Levy, David. *The Sky: A User's Guide.* Cambridge: Cambridge University Press, 1991. *An outstanding overview of various astronomical projects for the amateur; Levy's writing style is always enjoyable.*

Mechler, Gary. *First Field Guide Night Sky.* New York: Scholastic, 1999. *An introductory guide to the night skies, under the auspices of the National Audubon Society.*

Ottewell, Guy. *Astronomical Calendar.* Greenville, S.C.: Furman University, Universal Workshop, 1974–.*This outstanding annual guide with lavish illustrations covers each month's activities in detail, including a section on meteor showers.*

Seeds, Michael A. *The Solar System.* Belmont, Calif.: Wadsworth Publishing, 1999. *A popular college text that will provide the viewer with a solid solar system overview. Text also includes a CD-ROM for both Windows and Macintosh.*

Van Holt, Tom. *Stargazing*. Mechanicsburg, Pa.: Stackpole Books, 1999.
   An introduction to observing the night sky, along with mythology; nicely
   illustrated.

PERIODICALS

The following monthly periodicals review upcoming meteor showers, as well as report on shower activity. Important meteorite falls are often covered.

Astronomy
21027 Crossroads Circle
Waukesha, WI 53187
800-446-5489
Email: customerservice@kalmbach.com
www.kalmback.com/astro/astronomy

Mercury
Astronomical Society of the Pacific
390 Ashton Avenue
San Francisco, CA 94112
www.aspsky.org

The Planetary Report
The Planetary Society
65 North Cataline Avenue
Pasadena, CA 91106
www.planetary.org

Sky & Telescope
Post Office Box 9111
Belmont, MA 02178-9111
800-253-0245
Email: orders@skypub.com
www.skypub.com

The Strolling Astronomer
The Association of Lunar and Planetary Astronomers
Post Office Box 171302
Memphis, TN 38318-1302
www.lpl.arizona.edu/alpo

The following periodicals about meteorites are available:

Meteorite!
Pallasite Press
Post Office Box 33-1218
Takapuna, Auckland
New Zealand
Phone: 9-486-2428, Fax: 9-489-6750
Email: j.schiff@auckland.ac.nz
www.meteor.co.nz

Voyage!
Post Office Box 9284
Providence, RI 02940-9284
401-353-4098
Fax: 401-353-7901
Email: voyagemag@aol.com

INTERESTING WEBSITES

www.meteoritecentral.com
> Meteorite information, meteorite bulletin board list

www.arachnaut.org/meteor/links
> A hypertext link to various meteorite dealers, papers on meteors and meteorites, organizations, etc.

Appendix B:
Organizations

The following list includes many organizations and groups worldwide that coordinate, sponsor, and reduce data from meteor observing.

AMERICAN METEOR SOCIETY

Dr. Charles P. Olivier founded the American Meteor Society (AMS) in 1911. The AMS supports beginning visual observers of meteors by providing brief observing directions, charts, forms, and other materials upon receipt of initial dues payment.

In recent years, advances in technology have made it possible for amateur meteor observers with fairly modest equipment to achieve professional-quality results, and it is in these areas that the AMS is concentrating its main efforts. The AMS desires to help coordinate amateur-professional activities among observers who are too widely scattered to permit the usual close working relationships found in local groups.

The AMS is primarily interested in visual, CCD camera, and radio observations of sporadic meteor rates, meteor shower rates, and meteor magnitudes as well as other meteor property information.

The AMS publishes several documents of interest:
- *Meteor Trails* is the AMS quarterly journal containing technical articles, observing reports and analyses, and information on upcoming meteor showers.
- *AMS Annual Report* is a compendium of activity reports and the activities of the AMS.

- *Special Bulletins* covering particular topics are occasionally published by the AMS.

Annual dues are $6.50 for students and observers, $8.00 for associates, and $10.00 for groups.

Contacts:
Karl Simmons
AMS Treasurer
3859 Woodland Heights
Callahan, FL 32011

Robert Lunsford
AMS Visual Coordinator
161 Vance Street
Chula Vista, CA 91910-4828

The AMS web site provides continually updated weekly meteor outlooks, an on-line form to report fireballs, reports made by other meteor observers, and general historical information about the AMS.
www.amsmeteors.org

ASSOCIATION OF LUNAR AND PLANETARY OBSERVERS
The Association of Lunar and Planetary Observers supports a meteor observing section.

Robert Lunsford
ALPO Meteors Section Recorder
161 Vance Street
Chula Vista, CA 91910
Email: lunro.imo.usa@home.com
www.lpl.arizona.edu/alpo

BRITISH ASTRONOMICAL ASSOCIATION
The British Astronomical Association (BAA) was founded in 1890 and is open to all people interested in astronomy. The BAA supports an active meteor observing section.

Contact Neil Bone by email at neil@bone2.freeserve.co.uk
www.ast.cam.ac.uk

DUTCH METEOR SOCIETY

The Dutch Meteor Society (DMS) is best known for photographic results and pioneering work in meteor astronomy. The DMS is the world's largest producer of precision photographic and videographic orbital elements. DMS members are also active in visual, fireball, and radar observations, as well as some meteorite research. The DMS publishes a bimonthly journal, the *Radiant*.

Dutch Meteor Society
Lederkarper 4
NL-2318 NB
Leiden, The Netherlands
Phone: +(31)-71-5223817
Fax: +(31)-71-5223817
Email: betlem@strw.leidenuniv.nl
www.dmsweb.org

INTERNATIONAL METEOR ORGANIZATION

The International Meteor Organization (IMO) was established in 1988 in response to the ever-growing need for international meteor amateur work collaboration. The IMO accepts observations and has compiled a database of meteor observations.

North American Contact (North American Meteor Network):
Mark Davis
101 Margate Circle
Goose Creek, SC 29445
Email: meteorobs@charleston.net
web.infoave.net/~meteorobs
www.imo.net

THE METEORITICAL SOCIETY

The Meteoritical Society is an international nonprofit organization devoted to the study of meteorites and other samples of extraterrestrial matter and their relation to the origin and history of the solar system. Areas of interest include asteroids, comets, craters, interplanetary dust, interstellar medium, lunar samples, meteorites, meteors, natural satellites, planets, and tektites.

Membership in the society includes a subscription to *Meteoritics and Planetary Science*. Members may also subscribe to the journal *Geochimica Cosmochimica Acta* at a reduced rate.

Annual dues are $90 for regular members, and $45 for students. To apply for membership, either download, complete, and mail a paper copy of the application form, or contact Dr. Greg Herzog and request a membership form.

The Meteoritical Society
Gregory Herzog, Treasurer
Department of Chemistry
Rutgers University,
610 Taylor Road
Piscataway, NJ 08854
Email: herzog@rutchem.rutgers.edu
www.uark.edu/campus-resources/metsoc

Appendix C:
Meteorites and Museums

There are several excellent museums and science centers as well as universities where you can see meteorites and related exhibits. Occasionally these facilities will present lectures on meteorites. In addition, many planetaria have nice meteorites on display; this listing is not to be considered extensive or complete.

AMERICAN MUSEUM OF NATURAL HISTORY
The American Museum of Natural History, the recently opened Rose Center for Earth and Space, and the Hayden Planetarium have extensive collections of meteorites on display, including several famous meteorites.

West 81st Street
New York, New York
212-769-5200
www.amnh.org/rose

CHABOT SPACE & SCIENCE CENTER
The new Chabot Space & Science Center collaborated with the Smithsonian Institution's National Museum of Natural History as an Affiliate to develop a meteorite exhibit and mural.

10000 Skyline Boulevard
Oakland, California
510-336-7600
Email: info@chabotspace.org
www.chabotspace.org

UNIVERSITY OF ARIZONA MINERAL MUSEUM
The Mineral Museum displays examples of meteorites and tektites from around the world.

Tucson, Arizona
520-621-4515
www.geo.arizona.edu/minmus

DENVER MUSEUM OF NATURE AND SCIENCE
2001 Colorado Boulevard
Denver, Colorado
303-322-7009 or 800-925-2250
www.dmnh.org

GRIFFITH OBSERVATORY
Griffith Observatory exhibits a fine collection of meteorites and tektites. Griffith houses one of the country's outstanding astronomy facilities including exhibits, a planetarium, and observatories.

Griffith Park
2800 East Observatory Road
Los Angeles, California
323-664-1191
www.griffithobs.org

METEORITE MUSEUM AT
THE INSTITUTE OF METEORITICS
Self-guided tours of a number of the over 550 meteorites in the University's collection.

University of New Mexico
Albuquerque, New Mexico
505-277-1643
www.unm.edu

METEOR CRATER

Located about five miles south of Interstate 40, twenty miles west of Winslow, and thirty-five miles east of Flagstaff, Meteor Crater and the Museum of Astrogeology offer the visitor a view of the best-preserved crater on earth. Numerous meteorites, including a 1,406-pound meteorite from the Meteor Crater impact, are on display. In addition, there are graphics regarding the crater's formation, other meteorites from other falls and finds, space program artifacts, movies, lectures, and a short crater rim guided tour.

Interstate 40-Exit 233 at Meteor Crater Road
Arizona
520-289-2362 or 800-289-5898
Email: info@meteorcrater.com
www.meteorcrater.com

PEABODY MUSEUM OF NATURAL HISTORY

The oldest meteorite collection in the United States, begun in 1807, containing thousands of meteorites and tektites.

Yale University
170 Whitney Avenue
New Haven, Connecticut
203-432-5050
www.peabody.yale.edu/collections/met

SMITHSONIAN INSTITUTION NATIONAL MUSEUM OF NATURAL HISTORY

As the nation's keeper of meteorite specimens, the National Museum of Natural History has an exquisite gem and mineral exhibit (Janet Annenberg Hooker Hall of Geology, Gems, and Minerals), including a meteorite exhibit hall.

10th and Constitution Avenue (accessible entrance)
Washington, D.C.
202-357-2700
www.nmnh.si.edu
NMNH National Meteorite Collection division URL:
    www.nmnh.si.edu/minsci/meteor

THE FIELD MUSEUM
The Field Museum's exhibit represents an established and respected meteorite collection of nearly 1,400 falls.

Chicago's Museum Campus
1400 South Lake Shore Drive
Chicago, Illinois
312-922-9410
www.fmnh.org

UNIVERSITY OF CALIFORNIA AT LOS ANGELES
One of the largest collections in the United States and the largest on the West Coast, containing about 1,600 specimens representing more than 650 different meteorites.

Department of Earth and Space Sciences
Geology Building Third Floor

Appendix D:
Meteorite Dealers

There are numerous meteorite dealers in the United States and overseas. This list is meant to be only a starting point and is not meant to exclude any particular dealer or endorse another. You should also do a web search for dealers, if possible.

At the annual Tucson Gem and Mineral Show in February, rock, mineral, fossil, gem, and, yes, meteorite dealers literally take over several hotels and each hotel room becomes a dealer's showcase. It is an excellent opportunity to meet many of the dealers, look at meteorite samples, and price meteorites.

Astronomical Research Network
206 Bellwood Avenue
Maplewood, MN 55117
651-488-5178
Email: arni@visi.com

Dean Bessey's Meteorite Shop
Email: themeteoriteshop@yahoo.com
www.meteoriteshop.com

Bethany Sciences
Post Office Box 3726
New Haven, CT 06525
203-393-3395

Michael Blood Meteorites
6106 Kerch Street
San Diego, CA 92115
619-286-4837
Email: mblood@access1.net
www.meteorite.com/Michael_Blood

Cosmic Cutlery
Bud Eisler
3906 West Ina Road
PMB 285
Tucson, AZ 85741
520-682-7542

Excalibur Mineral Company
1000 North Division Street
Peekskill, NY 10566
914-739-1134
www.bestweb.net/~excalmin

Mike Farmer
1001 W. St Marys #813
Tucson, AZ 85745
520-206-9153
Email: farmerm@concentric.net
www.concentric.net/~Farmerm

Fernlea Meteorites UK
Fernlea, The Wynd
Milton of Balgonie
Fife KY7 6PY United Kingdom
Email: fernlea4@aol.com

Robert A. Haag
Post Office Box 27527
Tucson, AZ 85726
520-882-8804

Fax: 520-743-7225
Email: bobhaag@primenet.com
www.meteoriteman.com

R. N. Hartman
Post Office Box 94
Walnut, CA 91788-0094
909-595-8533
www.meteorite1.com

Tim Heitz Meteorites
636-225-9449
Email: midwestmeteor@earthlink.net
www.meteorites.simplenet.com

International Meteorite Brokerage
Post Office Box 82
Kingston, AR 72742
501-665-2345
www.meteoritebroker.com

Island Meteorite
Geoffrey Cintron
516-731-8218
Email: geoffcw@aol.com
www.islandmeteorite.com

Jensen Meteorites
8503 W Mountain View Lane
Littleton, CO 80125
Email: mikestockj@aol.com
www.meteorite.com/Jensen

Labenne Meteorites S.A.R.L.
Email: metlab.1@worldnet.fr
www.labenne-meteorites.com

RA Langheinrich Meteorites & Fossils
290 Brewer Road
Ilion, NY 13357
Phone and Fax: 732-764-0879
Email: meteorite@compuserve.com
www.nyrockman.com

The Macovich Collection of Meteorites
Darryl Pitt
1501 Broadway
Suite 1304
New York, NY 10036
212-302-9200
Fax: 212-382-1639
Email: darrylpitt@ren.com
www.macovich.com

Mare Meteoritics
Mike Martinez
Post Office Box 19041
Oakland, CA 94619
Phone and Fax: 925-743-8146
Email: meteors@flash.net
www.flash.net/~meteors

The Meteorite Market
907-789-6800
Email: twelker@alaska.net
www.meteoritemarket.com
www.alaska.net/~meteor

Meteorites.com
Email: info@meteorites.com
www.meteorites.com

Mile High Meteorites
Matt Morgan
Email: mhmeteorites@yahoo.com
www.mhmeteorites.com

Mineralogical Research Co.
Eugene and Sharon Cisneros
15840 East Alta Vista Way
San Jose, CA 95127-1737
408-923-6800
Fax: 408-926-6015
Email: xtls@minresco.com
www.minresco.com

New England Meteoritical Services
Post Office Box 440
Mendon, MA 01756
508-478-4020
Fax: 508-478-5104
Email for catalog mailing list: mailings@touchanotherworld.com
Email for ordering a meteorite: orders@touchanotherworld.com
www.meteorlab.com
www.touchanotherworld.com

Bill Peck
Box 771559
Steamboat Springs, CO 80477
970-879-3621
Email: bpeck@meteoritemap.com

Blaine Reed
Post Office Box 1141
Delta, CO 81416
Phone/Fax: 970-874-1487

The Meteorite Exchange, Inc.
Post Office Box 7000
Redondo Beach, CA 90277-8710
www.meteorite.com
This URL is a direct link to numerous meteorite dealers.

Riker Specimen Mounts are available from the following sources:

Carolina Biological Supply Company
2700 York Road
Burlington, NC 27215
800-227-1150
Fax: 800-222-7112
www.carolina.com

V Rock Shop
7061 Sunset Strip Avenue NW
Canton, OH 44720-7078
330-494-1759
Fax: 330-494-1432
Email: ohio@rockshop.com
www.rockshop.com

Bug boxes for micrometeorite collections are available from:

Alpha Supply
Post Office Box 2033
Bremerton, WA 98310
800-257-4211
Fax: 800-257-4244

Appendix E:
Meteorite Verification Laboratories

Do you have a specimen you have found and would like to have confirmed as a meteorite? You should first contact your local science museum, planetarium, or observatory. Additionally, you can contact one of the following institutions for assistance. Send your specimen with return receipt requested (some also insure their "find") and include postage for return of the specimen.

The following information should be included with your suspect meteorite:

- Date of find, including how you found the specimen. (For example, was this an observed fall?)
- Location of find (GPS coordinates, if possible).
- Description of how the specimen and ground appeared (evidence of impact?); photographs are best.

Institute of Geophysics and Planetary Sciences
University of California at Los Angeles
Los Angeles, CA 90024

Chabot Space & Science Center
Attention: Dr. Mike Reynolds
10000 Skyline Boulevard
Oakland, CA 94619

The American Museum of Natural History
Central Park West at 79th Street
New York, NY 10024

Center for Meteorite Studies
Arizona State University
Tempe, AZ 85281

Lunar and Planetary Institute
Space Sciences Building
University of Arizona
Tucson, AZ 85721

National Museum of Natural History
Department of Mineral Sciences
Smithsonian Institution
Washington, D.C. 20560

Appendix F:
Sample Meteorites

Below you will find a selection of thirty meteorites, including the location of the find or fall, general information about each meteorite, as well as a pricing guide. These meteorites represent some of the better-known samples, many of which are considered outstanding specimens.

ABEE, ALBERTA, CANADA
Latitude 54° 13' N, Longitude 113° 0' W
Stone; Enstatite Chondrite (EH)
Prices: 1989-$30.63/gram; 2000-$50.00/gram

This was a fall witnessed in 1952. A 107-kilogram stone was recovered from a six-foot-deep hole.

ALLENDE, CHIHUAHUA, MEXICO
Latitude 26° 58' N, Longitude 105° 19' W
Stone; Carbonaceous Chondrite (CV3.2)
Prices: 1984-$2.90/gram; 2000-$10.00/gram

A literal shower of stones fell after a bright bolide was seen on the morning of February 8, 1969. More than two tons of meteorites have been collected; probably another two tons are still in the area.

BARRATTA, DENILIQUIN, COUNTY TOWNSEND, NEW SOUTH WALES, AUSTRALIA
Latitude 35° 18' S, Longitude 144° 34' E
Stone; Ordinary Chondrite (L3.8)
Prices: 1988-$2.81/gram; 2000-$6.00/gram

The first stone was discovered in 1845. In all, five stones, weighing approximately 79, 66, 22, 22, and 14 kilograms, were found at different times.

BOXHOLE, NORTHERN TERRITORY, AUSTRALIA
Latitude 22° 37' S, Longitude 135° 12' E
Iron; Medium Octahedrite (IIIAB)
Prices: 1982-$1.49/gram; 2000-$4.00/gram

A circular depression near Huckitta, about 558 feet in diameter, was recognized as a meteorite crater in 1937. It is not known exactly how much material has been found in connection with the crater; an estimate, based on various reports and collections, would be that at least five hundred kilograms of iron meteorites and perhaps fifty kilograms of shale balls have been recovered. The major part of the main mass probably vaporized or was disseminated as minute melted globules.

BRAHIN, MINSK, BELORUSSIYA
Latitude 52° 30' N, Longitude 30° 20' E
Stony-Iron; Pallasite (PAL)
Prices: 1993-$7.58/gram; 1999-$30.00/gram

This meteorite was first discovered in 1810. Two masses weighing about 80 and 20 kilograms were recovered that year. A third mass of 183 kilograms was later recovered. There has been considerable confusion over the number and weights of the known pieces of this fall; up to eleven pieces have been reported.

CAMEL DONGA, NULLARBOR PLAIN, WESTERN AUSTRALIA

Latitude 30° 19' S, Longitude 126° 37' E
Stone; Achondrite, Eucrite (EUC)
Prices: 1990-$8.48/gram; 2000-$50.00/gram

This meteorite, found in 1984, displays a glossy black fusion crust. Over ten kilograms of this rare type of meteorite, which is composed mainly of pyroxene and plagioclase, were recovered in the desert of the Nullarbor Plain. Although no one is certain when this meteorite fell, specimens are not very weathered and appear to have fallen rather recently.

CANYON DIABLO, METEOR CRATER, COCONINO COUNTY, ARIZONA

Latitude 35° 3' N, Longitude 111° 2' W
Iron; Coarse Octahedrite (IAB)
Prices: 1984-31¢/gram; 2000-$1.00/gram

This crater and associated meteorites were first noted in 1891. Specimens contain microscopic diamonds. Estimates of the original mass are five thousand to twenty-five thousand tons. Collecting is forbidden at the crater site.

ESQUEL, CHUBUT ARGENTINA

Latitude 42° 54' S, Longitude 71° 20' W
Stony-Iron; Pallasite (PAL)
Prices: 1992-$2.00/gram; 2000-$35.00/gram

This meteorite, a single mass of fifteen hundred kilograms, was found prior to 1951. Meteorite dealer Robert Haag purchased the entire mass.

GIBEON, GREAT NAMA LAND, NAMIBIA, AFRICA

Latitude 25° 30' S, Longitude 18° E
Iron; Fine Octahedrite (IVA)
Prices: 1990-90¢/gram; 2000-$2.00/gram

Found in 1836 but known long before that. Natives used Gibeon meteorites as a source of iron. Gibeon meteorites are often used today to make collectors knives, wedding rings, and other ornamental items also considered collectibles.

GREAT BEND, BARTON COUNTY, KANSAS
Latitude 38° 23' 42" N, Longitude 98° 54' 57" W
Stone; Ordinary Chondrite (H6)
Prices: 1985-$3.00/gram; 1996-$3.96/gram

This meteorite was discovered in 1983. Only one single mass, weighing 28.77 kilograms, was found. This meteorite has a dark brown stone matrix, with small metallic fragments throughout.

GRETNA, PHILLIPS COUNTY, KANSAS
Latitude 39° 56' N, Longitude 99° 13' W
Stone; Ordinary Chondrite (L5)
Prices: 1988-$2.00/gram; 1997-$1.39/gram

In 1912, two pieces of a broken stone were recovered about 12 miles north of Gretna, Kansas. The two pieces weigh 36 kilograms and 22.7 kilograms and fit together, indicating the loss of a third fragment, probably about 23 kilograms in weight. The total meteorite probably weighed about 82 kilograms.

HAPPY CANYON, ARMSTRONG COUNTY, TEXAS
Latitude 34° 48' 6" N, Longitude 101° 34' W
Stone; Enstatite Chondrite (EL)
Prices: 1989-$4.98/gram; 1999-$95.00/gram

In 1971, a single mass of about 16.3 kilograms was ploughed up about 1.2 miles northwest of Wayside, Texas. The meteorite was cut and one section removed; the whereabouts of this section is not known.

HAXTUN, PHILLIPS COUNTY, COLORADO
Latitude 40° 27' 4" N, Longitude 102° 34' 7" W
Stone; Ordinary Chondrite (L4)
Prices: 1994-$4.97/gram; 2000-$8.32/gram

In August of 1975, a single weathered stone weighing 15.5 kilograms was plowed up in a wheat field.

HENBURY, NORTHERN TERRITORY, AUSTRALIA
Latitude 24° 34' S, Longitude 133° 10' E
Iron; Medium Octahedrite (IIIAB)
Prices: 1990-$0.60/gram; 2000-$2.00/gram

Shortly after discovery, large quantities of this meteorite were removed from the area, and over the years the site has been almost completely stripped of its fragments. As a result, the government has closed most of the area and collecting is forbidden. It has been estimated that at least twelve hundred kilograms of iron meteorite fragments have been recovered from the crater field, most of the material having been collected from sites outside the craters.

HOBA, GROOTFONTEIN, NAMIBIA
Latitude 19° 35' N, Longitude 17° 55' E
Iron; Ataxite, Nickel-rich (IVB)
Prices: 1991-$7.50/gram; 1999-$41.34/gram

In 1920 a mass of very ductile iron, measuring over 9 by 9 by 3.2 feet, was found on a farm twelve miles west of Grootfontein. The weight of the meteorite is estimated from measurements to be at least sixty metric tons. The main mass remains at the place of fall.

HOLBROOK, NAVAHO COUNTY, ARIZONA

Latitude 34° 54' N, Longitude 110° 11' W
Stone; Ordinary Chondrite (L6)
Prices: 1990-$5.00/gram; 2000-$10.00/gram

A witnessed fall on July 12, 1912 from a bright fireball. Over 14,000 stones fell, varying in size from sand up to a big grapefruit. Holbrook's estimated total mass is about eleven hundred kilograms.

IMILAC, ATACAMA DESERT, CHILE
Latitude 24° 12' S, Longitude 68° 48' W
Stony-Iron; Pallasite (PAL)
Prices; Grade 1: 1989-$7.44/gram; 2000-$35.00/gram
Prices; Grade 2: 1986-$2.95/gram; 2000-$10.00/gram

This rather famous pallasite was first discovered in 1882. Grade 1 specimens contain beautiful olivine crystals, and Grade 2 specimens are smaller pallasites without the olivine crystals.

INGELLA STATION, SOUTH GREGORY, QUEENSLAND, AUSTRALIA

Latitude 25° 33' S, Longitude 142° 47' E
Stone; Ordinary Chondrite (H5)
Prices: 1989-$5.00/gram; 1994-$2.00/gram

In 1987, a weathered stone weighing about ten kilograms was found in the Tenham area. It was recognized as a separate new find within this strewn field. The Ingella Station meteorite was found in a very weathered state, with traces of old fusion crust remaining. A large mass of approximately ten kilograms and a second of seven kilograms were found by searching the nearby fields. Additional meteorites were later recovered.

MILLBILLILLIE, WILUNA DISTRICT, WESTERN AUSTRALIA
Latitude 26° 27' S, Longitude 120° 22' E
Stone; Achondrite Eucrite (EUC)
Prices: 1989-$3.85/gram; 2000-$30.00/gram

A witnessed fall in October 1960; first recovered in 1970. Total recovered weight of about twenty-five kilograms. Millbillillie is a rare meteorite and is suspected to be from the asteroid Vesta.

MURCHISON, VICTORIA, AUSTRALIA
Latitude 36° 37' S, Longitude 145° 12' E
Stone; Carbonaceous Chondrite (CM2)
Prices: 1988-$16.96/gram; 2000-$100.00/gram

On a Sunday morning, September 28, 1969, between 10:45 and 11:00 A.M., in Murchison, Australia, a fireball exploded, signaling the arrival of the Murchison meteorites. Loud explosions and hissing noises were heard throughout the area, and smoke rings were seen hanging in the air. The meteorites were scattered across a twenty square mile area, stinking up the town of Murchison with the smell of methylated spirits and dust. Looking more like unburned charcoal briquettes than rocks, people gathered the meteorites from their yards and neighborhood streets. Some seven hundred kilograms of stones rained down out of the sky, the largest weighing only about seven kilograms.

NAKHLA, ABU HOMMOS, ALEXANDRIA, EGYPT

Latitude 31° 19' N, Longitude 30° 21' E
Stone; Achondrite Nakhlite (SNC)
Prices: 1993-$517/gram; 1999-$3,500/gram

About forty stones with a total weight of about 40 kilograms, varying from 20 grams to 1,813 grams, fell June 28, 1911. This rare and highly desirable meteorite is believed to have originated from the planet Mars, thus is called a "Mars rock." One of the stones from the Nakhla fall was reported to have killed a dog.

NANTAN, GUANGXI, CHINA
Latitude 25° 6' N, Longitude 107° 42' E
Iron; Medium Octahedrite (IIICD)
Prices: 1992-$2.00/gram; 2000-10¢/gram

At least nineteen specimens with a mass of approximately 9,500 kilograms were discovered in 1958. Some of these specimens impacted in an area below the water table and are "rusters." Do not cut these specimens; Nantan meteorites disintegrate!

NEW CONCORD, MUSKINGUM COUNTY, OHIO
Latitude 40° 0' N, Longitude 81° 46' W
Stone; Ordinary Chondrite (L6)
Prices: 1988-$4.08/gram; 1998-$15.00/gram

On May 1, 1860 at 12:45 P.M., an explosive fireball distributed about thirty stones over an area of about ten miles by three miles. The largest stone weighed 46.7 kilograms and the total weight recovered was about 227 kilograms. One of the falling stones struck and killed a colt.

ODESSA, ECTOR COUNTY, TEXAS
Latitude 31° 43' N, Longitude 102° 24' W
Iron; Coarse Octahedrite (IAB)
Prices: 1990-16¢/gram; 2000-$1.00/gram

Discovered in 1923, the impact was composed of thousands of individual meteorites of various sizes that fell over an area of about two square miles. The smaller meteorites, which were by far the most numerous, came to rest on the earth's surface or at the bottom of shallow impact pits within the soil.

OZONA, CROCKET COUNTY, TEXAS
Latitude 30° 44' N, Longitude 101° 18' W
Stone; Ordinary Chondrite (H6)
Prices: 1984-$1.50/gram; 2000-$3.00/gram

A rancher discovered this weathered chondrite on Mrs. Payen's Ranch in 1929. It was recognized as a meteorite in 1939, then sold to the Field Museum in Chicago. Later it was traded back to the El Paso Centennial Museum. Several additional fragments were recovered.

PLAINVIEW, HALE COUNTY, TEXAS
Latitude 34° 7' N, Longitude 101° 47' W
Stone; Ordinary Chondrite (H5), Brecciated
Prices: 1988-$4.49/gram; 2000-$1.99/gram

About twelve meteorites have been found since 1917. This fall was probably witnessed by the Norfleet family living about fifteen miles west-southwest of where most of the stones were found. A five-kilogram stone fell in the Norfleet horse corral and was dug out the next day. A study of this stone indicates that it may very well have been one of the heavier fragments that carried farther than the bulk of the shower, which was found scattered over an area of sixteen by four miles. This fall was in 1902 or 1903, and the area was sparsely settled at that time. The total weight recovered is over seven hundred kilograms.

SIKHOTE-ALIN, MARITIME TERRITORY, FEDERATED SSR
Latitude 46° 9' 36" N, Longitude 134° 39' 12" E
Iron; Coarsest Octahedrite (IIAB)
Prices, Shrapnel: 1984-$2.50/gram; 1999-$1.00/gram
Prices, Fusion Crust: 1993-$2.50/gram; 2000-$5.00/gram

The largest shower witnessed in historical time occurred in Eastern Siberia on February 12, 1947. In full daylight, a fireball moved from north to south and at about 10:38 A.M. local time fragmented in the earth's atmosphere. The brightness exceeded that of the sun, according to eyewitnesses. When this meteorite fell it literally went off like a grenade. The dust trail was observed for several hours before the particles precipitated or were scattered by the wind. The debris covered an elliptical area of about a mile on the snow-covered western spurs of the Sikhote-Alin Mountains, leaving over 120 craters. Most of the impacting meteorites did not penetrate the debris. It appears plausible that the incoming bolide had a mass of about seventy tons.

TOLUCA, MEXICO STATE, MEXICO
Latitude 19° 34' N, Longitude 99° 34' W
Iron; Coarse Octahedrite (IAB)
Prices: 1993-$0.19/gram; 2000-$1.85/gram

The important Toluca shower comprises many thousands of fragments that have been recovered from a rather small area on the hillsides around the village of Xiquipilco, which is situated in a remote branch of the Toluca Valley network. The cumulative weight of all specimens in registered collections is about 2,100 kilograms. To this must be added numerous fragments that were forged into agricultural implements. The Indians of Xiquipilco make spades and axes of the iron, and the owners of the haciendas use it for plows. The largest known individual is in Mexico City in the Institute of Geology. It weighs approximately four hundred kilograms.

VACA MUERTA, TALTAL DISTRICT, ATACAMA, CHILE
Latitude 25° 45' S, Longitude 70° 30' W
Stony-Iron; Mesosiderite (MES)
Prices: 1987-$2.25/gram; 2000-$5.00/gram

Large masses up to twenty-five kilograms were discovered in 1861 and originally found on a cattle trail along with cow bones, thus the name Vaca Muerta—"dead cow."

ZAGORA, MOROCCO, AFRICA
Latitude 32° 22' N, Longitude 5° 51' W
Iron; (IAB) with silicate inclusions
Prices: 1989-$3.90/gram; 2000-$25.00/gram

In 1987 fossil hunters found this meteorite in the desert near the Atlas Mountains. Unfortunately, the locality is near a military zone. At this time it is not possible to visit the site to search for additional meteorites.

Index